Nonlinear Optimization

THE INTERNATIONAL SERIES OF MONOGRAPHS ON COMPUTER SCIENCE

1. *The design and analysis of coalesced hashing*
   Jeffrey S. Vitter and Wen-chin Chen
2. *Initial computability, algebraic specifications, and partial algebras*
   Horst Reichel
3. *Art gallery theorems and algorithms*
   Joseph O'Rourke
4. *The combinatorics of network reliability*
   Charles J. Colbourn
5. *Discrete relaxation techniques*
   T. Henderson
6. *Computable set theory*
   D. Cantone, A. Ferro, and E. Omodeo
7. *Programming in Martin-Löf's type theory: an introduction*
   Bengt Nordström, Kent Petersson, and Jan M. Smith
8. *Nonlinear optimization: Complexity Issues*
   Stephen A. Vavasis

NONLINEAR OPTIMIZATION

Complexity Issues

STEPHEN A. VAVASIS
Cornell University

New York Oxford
OXFORD UNIVERSITY PRESS
1991

Oxford University Press

Oxford New York Toronto
Delhi Bombay Calcutta Madras Karachi
Petaling Jaya Singapore Hong Kong Tokyo
Nairobi Dar es Salaam Cape Town
Melbourne Auckland

and associated companies in
Berlin Ibadan

Published by Oxford University Press, Inc.,
200 Madison Avenue, New York, New York 10016

Oxford is a registered trademark of Oxford University Press

Library of Congress Cataloging-in-Publication Data
Vavasis, Stephen A.
Nonlinear optimization :
complexity issues / Stephen A. Vavasis.
p. cm.
Includes bibliographical references and index.
ISBN 0-19-507208-1
1. Computational complexity. 2. Mathematical optimization.
3. Nonlinear theories. I. Title.
QA267.7.V38 1991 519.3—dc20 91-17398

2 4 6 8 9 7 5 3 1

Printed in the United States of America
on acid-free paper

To my family
Andrew, Theana,
Valentina and Anthony

PREFACE

Optimization entered the modern era with Dantzig's development of the simplex method in the 1940s. The field of computer science also has its origins during that period. The fields of computer science and optimization have grown together and greatly influence each other. This book is about one important connection: complexity theory.

Complexity theory is the theory underlying computer algorithms. How efficient is an algorithm? Does an efficient algorithm exist for a certain problem? These are the questions that complexity theory addresses. Complexity theory originated in the '60s with work by Hartmanis and Stearns and was connected to many important computational problems (including optimization) with the discovery of NP-completeness by Cook and Karp in the early '70s.

The first strong result about the complexity of continuous (i.e., linear and nonlinear) optimization was Khachiyan's 1979 result that the ellipsoid algorithm solves linear programming in polynomial-time in the Turing machine model. The most significant development in optimization during the '80s was the discovery of interior point methods, pioneered by Karmarkar. Complexity theory played an important role in this discovery: Karmarkar and other researchers were aware of the paradox at the time that there seemed to be no correspondence between the algorithms efficient in practice for linear programming (simplex-based methods) and the polynomial-time algorithms (ellipsoid-based methods). Interior point methods appear to possess both kinds of efficiency, thereby resolving this paradox.

I believe that complexity theory will play an ever greater role in the discovery of new optimization algorithms. Optimization problem sizes have steadily grown with available computer hardware, and this trend will no doubt continue. As increasingly larger problems are attempted, it is more important than ever to understand the asymptotic complexity issues.

This book is meant to describe some of the important developments in complexity aspects of optimization during the last 10 years. This book is not intended to be a practical guide to optimization: For practical information I recommend the excellent books *Practical Optimization* by Gill, Murray and Wright or *Practical Methods of Optimization* by Fletcher. Indeed, this present book could serve as a useful complement to those texts.

Another recommended book is Murty's 1988 book on complementarity problems. Murty addresses many complexity issues connected to complementarity. Because complementarity is closely related to optimization, this leads to an alternative viewpoint for some of the results in this book.

This book is based on the notes for a one-semester graduate course taught at Cornell University in the spring of 1990. Prerequisites for this material include linear algebra, calculus of several variables, and acquaintance with computer algorithms. In Chapters 1 and 2 we describe the necessary background in optimization and complexity theory, although the treatment is brief. The reader not already familiar with these areas might wish to review other texts on complexity and optimization. Recommended texts on optimization include the books mentioned in the previous paragraphs. Recommended texts on complexity theory would be *Computers and Intractability* by Garey and Johnson, *Design and Analysis of Computer Algorithms* by Aho, Hopcroft and Ullman, and *Combinatorial Optimization* by Papadimitriou and Steiglitz. I have been especially influenced by the latter book; for instance, NP in this book is defined in terms of certificates rather than nondeterministic Turing machines.

I want to thank Mike Todd of Cornell for a number of comments that improved the presentation. I also want to thank the Mathematical Sciences Institute of Cornell, funded by the U.S. Army Research Office, and the U.S. Department of Energy, for funding some of the research involved in preparing this book.

This book was prepared in camera-ready form by the author using L. Lamport's LaTeX document preparation system and D. Knuth's TeX typesetting system. TeX is a trademark of the American Mathematical Society. The index was prepared with Chen's MakeIndex package. All of the figures in this book were prepared by the author using a combination of PostScript™ and Matlab™. PostScript, a graphics language, is a trademark of Adobe Systems Inc. Matlab, an interactive language for numerical computation, is a trademark of The Mathworks, Inc. The fonts are Adobe Times-Expanded for text and D. Knuth's Computer Modern for mathematical formulas. The postscript assembly process was carried out by the dvips program by Radical Eye Software. The production was done on a Linotronic L300 machine by Cayuga Press of Ithaca.

Ithaca, New York S. A. V.
February, 1991

CONTENTS

1. Optimization and convexity, 1

1.1. Classes of optimization problems, 2
1.2. Mathematical preliminaries, 4
1.3. Local optimality, 5
1.4. Convexity and KKT conditions, 6
1.5. Quadratic programming and convexity, 12

2. Complexity theory, 17

2.1. Turing machines, 17
2.2. P and NP, 18
2.3. NP-completeness, 24
2.4. Models of computation, 33

3. Convex quadratic programming, 36

3.1. The HKL algorithm for convex knapsack, 37
3.2. An interior point method, 42
3.2.1. The main loop of the Renegar-Shub algorithm, 47
3.2.2. Finding a global minimum, 56
3.2.3. Initializing the algorithm, 62
3.2.4. Running time analysis, 68
3.2.5. Polynomial number of digits, 69
3.2.6. Strongly polynomial time, 73

4. Nonconvex quadratic programming, 76

4.1. QP is NP-complete, 76
4.2. Special cases of nonconvex QP, 80
4.3. Nonconvex quadratic functions on a sphere, 84
4.4. Algorithms for indefinite quadratic problems, 99

5. Local optimization, 103

5.1. General quadratic local minimization, 103
5.2. Characterizing local minima for quadratic programs, 104
5.3. Local minima for concave knapsack problems, 107
5.4. An efficient implementation of local minimization, 114
5.5. The indefinite case of QKP, 119

6. Complexity in the black-box model, 135

6.1. Global minimization, 136
6.2. Local minimization, 140
6.3. Convex minimization and the ellipsoid algorithm, 146

LIST OF FIGURES

1.1. Two convex subsets in two dimensions. 7
1.2. Two nonconvex subsets in two dimensions. 7
1.3. KKT conditions can fail to be necessary. 10
1.4. Plots of four quadratic functions of two variables. 15

2.1. A schematic of a Turing machine. 19
2.2. The graph produced from $x_1 \wedge (\neg x_1 \vee x_2) \wedge (\neg x_2 \vee x_3 \vee x_4)$. 28
2.3. An example graph for the subset sum transformation. 30

3.1. The function $h(\lambda)$ for the example. 42
3.2. A diagram indicating convergence of the main loop. 50

4.1. One and two dimensional simplices. 82
4.2. Model trust region when $n = 1$. 85
4.3. A level curve at the optimum point in two dimensions. 87
4.4. The three cases for $p(\mu)$. 90

5.1. Point $(0, -1)$ has no feasible descent directions. 108
5.2. Computation of ρ_i. 128

6.1. A function whose minimum can be hidden. 137
6.2. An impossible case for local minimization. 143
6.3. One step of the ellipsoid algorithm. 148
6.4. An example of U', $\mathbf{y}$ and $\mathbf{z}$ in the $n = 2$ case. 153

NOTATION

$\mathbb{R}$ The real numbers.

$\mathbf{a}, \mathbf{b}, \mathbf{c}, \ldots$ **and** $\boldsymbol{\alpha}, \boldsymbol{\beta}, \boldsymbol{\gamma}, \ldots$ Vectors over the real numbers.

$a, b, c, \ldots$ **and** $\alpha, \beta, \gamma, \ldots$ Scalars.

$A, B, C, \ldots$ Capital letters usually mean sets or matrices.

$f(\mathbf{x})$ Often stands for the objective function of an optimization problem, that is, a continuous map $\mathbb{R}^n \to \mathbb{R}$.

$\nabla f(\mathbf{x})$ Gradient (first derivative) of $f(\mathbf{x})$.

$\mathbf{x}^*$ Asterisk usually indicates a local or global optimum.

P, NP Complexity classes (see Chapter 2).

$\frac{1}{2}\mathbf{x}^T H\mathbf{x} + \mathbf{c}^T\mathbf{x}$ This is a quadratic objective function. Matrix H is assumed to be symmetric. Matrix H and vector $\mathbf{c}$ specify the function, and vector $\mathbf{x}$ is the argument.

$\mathbf{x}^T\mathbf{y}$ Inner product of two vectors.

$\lg x$ Logarithm base 2 of x.

$O(f(n))$ A function $g(n)$ is $O(f(n))$ if there exists c, n_0 such that $|g(n)| \leq cf(n)$ for all $n > n_0$.

$o(f(h))$ A function $g(h)$ is $o(f(h))$ if

$$\lim_{h \to 0} \frac{|g(h)|}{f(h)} = 0.$$

Nonlinear Optimization

1

OPTIMIZATION AND CONVEXITY

The main kind of problem under consideration in this book is the *nonlinear optimization* problem, sometimes called *nonlinear programming*. The problem is, given a set $D \subset \mathbb{R}^n$, and given a continuous function $f : D \to \mathbb{R}$, find

$$\min\{f(\mathbf{x}) : \mathbf{x} \in D\}$$

and the vector $\mathbf{x} \in D$ where the minimum is achieved. The word "nonlinear" means that function f is not a linear function, although we will also study linear problems in this book as a special case of nonlinear optimization. The phrase "nonlinear optimization" is also chosen to distinguish the problems under consideration from *combinatorial optimization* in which $\mathbf{x}$ is constrained to lie in a finite set. It would be hard to underestimate the importance of optimization to scientific computing. Most kinds of engineering design problems involve either linear or nonlinear optimization. These problems also occur in economics, game theory, planning and scheduling, physics, chemistry, probability and statistics.

It is possible to express practically any problem of mathematics as an optimization problem. Consider the following example, due to Murty [1988].

$$\begin{aligned} \text{minimize} \quad & (x^n + y^n - z^n)^2 + \sin^2 \pi n + \sin^2 \pi x + \sin^2 \pi y + \sin^2 \pi z \\ \text{subject to} \quad & x \geq 1, y \geq 1, z \geq 1, n \geq 3 \end{aligned} \tag{1.1}$$

This optimization problem achieves a global minimum of 0 on its domain iff Fermat's last theorem is false.

Since we have no intention of addressing Fermat's last theorem in this text, we limit the scope of our inquiry to special classes of optimization problems. The most important class for this book is quadratic programming. In the next section we define quadratic programming and other classes of problems under consideration. In Section 1.2 we address some of the mathematical preliminaries. In later sections we provide background on local optimality and convexity.

This book focuses on special cases of optimization that can be solved with efficient algorithms and on classes of problems that are computationally intractable. Other treatments of optimization such as Fletcher [1987] have a different goal, namely, to study algorithms which work well in practice for a broad range of problems.

There is a tradeoff in this regard: Algorithms of more general applicability are less likely to admit strong results about their efficiency. This book leans towards less generality and stronger efficiency results.

1.1. Classes of optimization problems

Two pieces of data define an optimization problem: f, called the *objective function*, and D, called the *feasible region*. In order to present a nonlinear optimization problem to a computer, it is necessary to encode f and D in a form accessible to the machine. There are several ways in which to carry this out. One way is to have f available as a subroutine that can be evaluated from other routines. Any other routine could produce a vector $\mathbf{x} \in \mathbb{R}^n$ and then call this subroutine to evaluate $f(\mathbf{x})$. It may be the case that $f(\mathbf{x})$ and $\nabla f(\mathbf{x})$ are both available as subroutines. The external-subroutine model is called a *black-box* model. This model occurs in the literature; for example, Dennis and Schnabel [1983] work mainly in this model of computation. Such a model allows very general problems to be addressed. We return to this topic in Chapter 6.

Similarly, D can be modeled as a black box. In particular, we would have a subroutine that would accept a vector $\mathbf{x} \in \mathbb{R}^n$ and return a yes/no answer, where "yes" indicates that $\mathbf{x} \in D$ and "no" indicates that $\mathbf{x} \notin D$. This model of computation for D occurs only rarely in the literature; because of its generality, not much progress can be made. Useful results can be demonstrated for some cases of the black-box model of a set—see, for example, Lovász [1986].

If $D = \mathbb{R}^n$, the optimization problem is said to be *unconstrained*. In this case, nothing else needs to be said about how to specify D since every vector is feasible.

Another way to specify f is to assume that it has a certain form, and to specify only the coefficients in that form. The two most common kinds of functions f are *linear objective functions,*

$$f(\mathbf{x}) = \mathbf{c}^T\mathbf{x}$$

where $\mathbf{c}$ is a vector in $\mathbb{R}^n$, and *quadratic objective functions,*

$$f(\mathbf{x}) = \frac{1}{2}\mathbf{x}^T H\mathbf{x} + \mathbf{c}^T\mathbf{x},$$

where H is an $n \times n$ real matrix and $\mathbf{c}$ is a vector. These objective functions are specified by numerical data, either by the vector $\mathbf{c}$ or by the pair $(H, \mathbf{c})$. Our notation for vectors and matrices is discussed further in the next section.

Note that in the expression $\frac{1}{2}\mathbf{x}^T H\mathbf{x}+\mathbf{c}^T\mathbf{x}$ we can always assume without loss of generality that H is symmetric. The reason is that $\mathbf{x}^T H\mathbf{x} = \mathbf{x}^T H^T\mathbf{x}$ so $\mathbf{x}^T H\mathbf{x} = \frac{1}{2}\mathbf{x}^T(H + H^T)\mathbf{x}$, and the matrix $(H + H^T)/2$ is symmetric.

Feasible region D can also be specified by assuming that it has a certain form. The most common kind of specification is by a sequence of *constraints*. These constraints are equations or inequalities involving $\mathbf{x}$. The most common kinds of constraints are linear constraints. In this case, D would be specified by a sequence of equations of the form

$$\mathbf{a}^T\mathbf{x} = b$$

or of inequalities of the form

$$\mathbf{a}^T\mathbf{x} \geq b$$

or

$$\mathbf{a}^T\mathbf{x} > b.$$

The latter kind of linear inequality is less common—see Section 1.2 for an explanation of why this kind of inequality is not useful for optimization problems. If D is specified as the set of vectors satisfying linear constraints, then the optimization problem is said to be *linearly constrained.*

A linearly constrained optimization problem with a linear objective function is called a *linear programming* (LP) problem. Linear programming has many applications including scheduling, planning, and flow computations. The subject has been actively studied since the simplex algorithm was discovered by Dantzig in 1947 (see Dantzig [1963]). In this book linear programming is treated as a special case of quadratic programming.

Quadratic programming (QP) refers to a linearly constrained optimization problem with a quadratic objective function. Quadratic programming has also been studied intensively in the last few decades. A number of general-purpose algorithms exist for this problem; see, for example, Murtagh and Saunders [1987].

Quadratic programming, as a generalization of linear programming, has all the same applications. In addition, there are classes of problems in planning and scheduling that can be expressed only with a quadratic objective function. Quadratic programming, because it is **NP**-hard (see Chapter 2), can also express most of the interesting combinatorial optimization problems. Finally, certain problems in game theory and economics are naturally expressed as quadratic programs.

Other kinds of constraints or ways of specifying D come up in practice. For example, D might be defined to include only those vectors whose entries have integer values. Such a problem is called an *integer program*. In example (1.1) the objective function includes terms to force all the variables to be integers. This problem could be expressed more succinctly as an integer program: Minimize $(x^n + y^n - z^n)^2$ subject to x, y, z, n integers, $n \geq 3$, $x, y, z \geq 1$. We do not discuss integer programming in this book, but it has been considered by many other authors.

1.2. Mathematical preliminaries

The set $\mathbb{R}^n$ is the usual n-dimensional Euclidean space. Its elements are normally regarded as column vectors, that is, $n \times 1$ matrices. This space has an *inner product*; the inner product of two vectors $\mathbf{x}$ and $\mathbf{y}$ is denoted by $\mathbf{x}^T\mathbf{y}$ and is a real number. This notation is matrix notation: The superscript T indicates the transpose of a matrix or vector. The Euclidean or 2-norm of a vector $\mathbf{x}$ is defined to be $(\mathbf{x}^T\mathbf{x})^{1/2}$ and is denoted by $\|\mathbf{x}\|_2$ or just $\|\mathbf{x}\|$. Another useful norm is $\|\mathbf{x}\|_\infty$, which for a vector $\mathbf{x} = (x_1, \ldots, x_n)^T$ is defined to be

$$\max(|x_1|, \cdots, |x_n|).$$

For more information on vector spaces, matrices, and norms, see Golub and Van Loan [1989].

Next we turn to topological definitions. Topology plays an important role in determining which kinds of optimization problems have solutions.

DEFINITION. *The* **open ball** B *of radius* $d > 0$ *centered at* $\mathbf{x}_c \in \mathbb{R}^n$ *is the set of points*

$$B = \{\mathbf{x} \in \mathbb{R}^n : \|\mathbf{x} - \mathbf{x}_c\| < d\}.$$

See Munkres [1975] for this definition and further background on the topological definitions of this section.

DEFINITION. *A subset* $E \subset \mathbb{R}^n$ *is* **open** *if for every point* $\mathbf{x} \in E$ *there is an open ball* B *of radius* $d > 0$ *centered at* $\mathbf{x}$ *and completely contained in* E.

EXAMPLES. The open interval (a, b) in $\mathbb{R}$ is an open set. An open ball in $\mathbb{R}^n$ is an open set. Finally, the region defined by strict inequalities $A\mathbf{x} > \mathbf{b}$ is an open set.

We remark that the notation $A\mathbf{x} > \mathbf{b}$ is interpreted as follows: A is an $m \times n$ matrix; $\mathbf{x} \in \mathbb{R}^n$ is the test variable for membership in D; and $\mathbf{b}$ is an m-vector. The inequality $A\mathbf{x} > \mathbf{b}$ is true if and only if the ith entry of the vector $A\mathbf{x}$ is greater than the ith entry of vector $\mathbf{b}$ for each $i = 1, \ldots, m$.

Optimization problems on open sets often have no solution. Consider, for example, the problem of minimizing $f(x) = 1 - x$ on the open interval $(0, 1)$. The greatest lower bound of the objective function is 0, but this bound is not achieved on the interval. Accordingly, optimization problems are most often posed on *closed* sets.

DEFINITION. *A subset* $D \subset \mathbb{R}^n$ *is* **closed** *if for any convergent sequence of points* $\mathbf{x}_1, \mathbf{x}_2, \ldots$ *lying in* D*, the limit point* $\mathbf{x}^*$ *of the sequence also lies in* D.

EXAMPLES. The closed interval $[a, b]$ is a closed set. The *closed ball* $\{\mathbf{x} \in \mathbb{R}^n : \|\mathbf{x} - \mathbf{x}_c\| \leq d\}$ is also closed under the preceding definition. The set of points satisfying the equations $A\mathbf{x} = \mathbf{b}$ is closed, as is the set $A\mathbf{x} \geq \mathbf{b}$. Finally, it can be proved that the complement with respect to $\mathbb{R}^n$ of any open set is closed, and vice versa.

Requiring a set D to be closed is not enough to guarantee that the optimization problem defined on D has a solution. For example, let D be the nonnegative real numbers, a closed set. The objective function $f(x) = 1 - x$ has no minimum on this set. This objective function takes on negative numbers of arbitrarily large magnitude on D. Such a function is said to be *unbounded.* An optimization problem is said to be *unbounded* if its objective function is unbounded on the feasible region. As another example, consider the function $f(x) = e^{-x}$ on the same domain. This function is bounded below by 0, but there is no point x such that $f(x) = 0$. These examples motivate the next two definitions.

DEFINITION. *A subset $D \subset \mathbb{R}^n$ is said to be* **bounded** *if D is contained in some closed ball (of finite radius).*

DEFINITION. *A subset of $D \subset \mathbb{R}^n$ is said to be* **compact** *if D is closed and bounded.*

THEOREM 1.1. *Let $f : D \to \mathbb{R}$ be a continuous function. Suppose D is nonempty and compact. Then f achieves its minimum on D, i.e., there is a point $\mathbf{x}^* \in D$ such that $f(\mathbf{x}^*) \leq f(\mathbf{x})$ for all $\mathbf{x} \in D$.* ∎

This theorem, which is proved in Munkres [1975] and other introductory topology books, makes compact sets particularly attractive for optimization problems. Optimization problems on compact sets are guaranteed to have solutions.

1.3. Local optimality

Consider the optimization problem: Minimize $f(\mathbf{x})$ subject to $\mathbf{x} \in D$ where D is a nonempty subset of $\mathbb{R}^n$. A point $\mathbf{x}^* \in D$ is said to be a *local minimum* for this problem if there exists an open set N containing $\mathbf{x}^*$ such that $f(\mathbf{x}^*) \leq f(\mathbf{x})$ for any $\mathbf{x} \in N \cap D$.

If $N = \mathbb{R}^n$ in this definition, then $\mathbf{x}^*$ is said to be a *global minimum.* "Global minimum" is another term for the solution of the optimization problem. Clearly any global minimum is also a local minimum, although the converse need not hold. Consider, for example, the function $f(x) = 4x - x^2$ on the interval $[0, 3]$. It can be verified that 3 is a local minimum for this function but not a global minimum.

Suppose a point $\mathbf{x}^* \in D$ is a local minimum for a function f. A vector $\mathbf{d} \in \mathbb{R}^n$ is said to be a *feasible direction* at $\mathbf{x}^*$ if there exists an $\alpha > 0$ such that $\mathbf{x}^* + t\mathbf{d} \in D$ for all $t \in [0, \alpha]$. A *descent direction* is a vector $\mathbf{d}$ such that there is an $\alpha > 0$ such that $f(\mathbf{x}^* + t\mathbf{d}) < f(\mathbf{x}^*)$ for all $t \in (0, \alpha)$. Finally, a vector $\mathbf{d}$ is called a *feasible descent direction* if it is both a feasible direction and a descent direction. Clearly, the presence of a feasible descent direction means that $\mathbf{x}^*$ is not a local minimum. On the other hand, the absence of feasible descent directions at a point $\mathbf{x}^*$ does not necessarily guarantee that $\mathbf{x}^*$ is a local minimum (see Chapter 5).

If f is differentiable at $\mathbf{x}^*$ and $\nabla f(\mathbf{x}^*)^T\mathbf{d} < 0$, then $\mathbf{d}$ must be a descent direction by definition of the derivative. Therefore, the statement that

$$\nabla f(\mathbf{x}^*)^T\mathbf{d} \geq 0 \tag{1.2}$$

for all feasible directions $\mathbf{d}$ is a necessary (but not sufficient) condition for the absence of feasible descent directions.

This condition is called the *first-order* necessary condition for local optimality. It is always satisfied at a local minimum of a differentiable function, and therefore at a global minimum. It is not in general sufficient for local minimality, however. Consider the function $f(x) = 4x - x^2$ on the interval $[0, 2]$. For this problem the first-order condition is satisfied at 2 even though 2 is not a local minimum (indeed, it happens to be a local maximum in this particular case). The condition fails at this point because $d = -1$ is a feasible descent direction but $f'(2) \cdot (-1) = 0$.

Necessary conditions for local minimality in terms of second derivatives can be formulated. In addition, there exist conditions sufficient for local minimality in terms of higher derivatives. There appear to be no simple conditions in terms of derivatives that are both necessary and sufficient for local minimality for a general nonlinear constrained problem.

We do not work out higher order conditions in this section. We use them for a specific problem in Chapter 4; they are derived for that particular problem in that chapter. We give second-order conditions for quadratic programming in Section 5.2.

1.4. Convexity and KKT conditions

Convexity plays a central role in the theory of optimization and is a recurrent theme in this book. The term "convex" can be applied either to a set or to a real-valued function. The definitions are as follows.

DEFINITION. *A set $D \subset \mathbb{R}^n$ is said to be* **convex** *if for any pair of points* $\mathbf{x}, \mathbf{y} \in D$ *and for any real number* $\lambda \in [0, 1]$ *the point*

$$(1 - \lambda)\mathbf{x} + \lambda\mathbf{y}$$

lies in D.

The singleton set and the entire space $\mathbb{R}^n$ are both convex. Further examples of convex sets are shown in Fig. 1.1. Examples of nonconvex sets are shown in Fig. 1.2. A convex set can be either bounded or unbounded, and can be open, closed, or neither.

A large body of theory has been developed to address convex sets; see, for example, Lovász [1986] or Grünbaum [1967].

FIG. 1.1. Two convex subsets in two dimensions.

FIG. 1.2. Two nonconvex subsets in two dimensions.

DEFINITION. *Let $D \subset \mathbb{R}^n$ be a convex set. A function $f : D \to \mathbb{R}$ is* **convex** *if the following inequality holds for any pair of points $\mathbf{x}, \mathbf{y} \in D$ and any real number $\lambda \in [0, 1]$:*

$$f((1-\lambda)\mathbf{x} + \lambda\mathbf{y}) \leq (1-\lambda)f(\mathbf{x}) + \lambda f(\mathbf{y}). \tag{1.3}$$

Some useful facts about convexity are as follows. These facts are left for the reader to prove in Exercise 1.1.

- The intersection of two convex sets is convex.
- The sum of two convex functions is convex.
- The pointwise maximum of two convex functions is convex.
- The function $f(\mathbf{x}) = \mathbf{a}^T\mathbf{x}$ is convex.
- If f is a convex function defined on a convex set E, then the set $D = \{\mathbf{x} \in E : f(\mathbf{x}) \leq a\}$ is convex for any choice of the real number a.

Much is known about convex functions; see, for example, Rockafellar [1970]. The most important results for our purposes are the following theorem and Theorem 1.5.

THEOREM 1.2. *Let D be a nonempty convex domain and let f be a convex function on D. Suppose $\mathbf{x}^*$ be a local minimum of f. Then $\mathbf{x}^*$ is also a global minimum.*

PROOF. Since $\mathbf{x}^*$ is a local minimum, there is an open set N containing $\mathbf{x}^*$ in which this point is optimal. Choose an arbitrary $\mathbf{x} \in D$; we want to show that $f(\mathbf{x}) \geq f(\mathbf{x}^*)$. Consider the sequence of points $\mathbf{x}^{(i)}$ given by

$$\mathbf{x}^{(i)} = (1 - (1/i))\mathbf{x}^* + (1/i)\mathbf{x}$$

for $i = 1, 2, 3, \ldots$ Since D is convex, $\mathbf{x}^{(i)} \in D$ for all i. Moreover, for i large enough, $\mathbf{x}^{(i)} \in N$ since this sequence converges to $\mathbf{x}^*$. Therefore, $f(\mathbf{x}^{(i)}) \geq f(\mathbf{x}^*)$ for all i large enough. Since f is convex, we calculate:

$$f(\mathbf{x}^{(i)}) \leq (1 - (1/i))f(\mathbf{x}^*) + (1/i)f(\mathbf{x})$$

so

$$f(\mathbf{x}^*) \leq (1 - (1/i))f(\mathbf{x}^*) + (1/i)f(\mathbf{x})$$

for i large enough. This simplifies to

$$(1/i)f(\mathbf{x}^*) \leq (1/i)f(\mathbf{x}).$$

Multiplying by i gives the result. ∎

The preceding theorem does not establish the existence of a global minimum. If the feasible set is convex but not compact, then a convex function need not have a minimum. The minimum of a convex function, even if it exists, is not necessarily unique. There is a condition for uniqueness that we call upon later. A function $f : D \to \mathbb{R}$ is *strictly convex* if for all $\mathbf{x}, \mathbf{y} \in D$ such that $\mathbf{x} \neq \mathbf{y}$, and for all $\lambda \in (0, 1)$,

$$f((1 - \lambda)\mathbf{x} + \lambda\mathbf{y}) < (1 - \lambda)f(\mathbf{x}) + \lambda f(\mathbf{y}).$$

THEOREM 1.3. *Let D be a convex domain, and let $f : D \to \mathbb{R}$ be strictly convex. Then f has at most one global minimum.*

PROOF. Let $\mathbf{x}^* \in D$ be a global minimum, and let $\mathbf{x}$ be a point in D different from $\mathbf{x}^*$. Let $\mathbf{x}_m = 0.5\mathbf{x} + 0.5\mathbf{x}^*$. By convexity, $\mathbf{x}_m \in D$. Since $\mathbf{x}^*$ is a global minimum, $f(\mathbf{x}^*) \leq f(\mathbf{x}_m)$. By strict convexity,

$$f(\mathbf{x}_m) < 0.5f(\mathbf{x}^*) + 0.5f(\mathbf{x}),$$

so

$$f(\mathbf{x}^*) < 0.5f(\mathbf{x}^*) + 0.5f(\mathbf{x}).$$

Simplifying shows that $f(\mathbf{x}^*) < f(\mathbf{x})$. ∎

Minimizing a convex function on a convex set comes up often in practice and has the strong property given by the two previous theorems. Accordingly, we give it a name.

DEFINITION. *The problem of minimizing a convex function $f(\mathbf{x})$ on a convex domain $D \in \mathbb{R}^n$ is called* **convex programming** *or* **convex optimization**.

In the previous section we gave a first-order necessary condition in terms of feasible directions. First-order conditions more amenable to computing can be derived as follows. Suppose D is a domain given by a sequence of constraints. These constraints have the following form:

$$\begin{aligned} r_1(\mathbf{x}) &\geq a_1 \\ &\vdots \\ r_s(\mathbf{x}) &\geq a_s \end{aligned} \tag{1.4}$$

where each function r_i is a real-valued differentiable function on $\mathbb{R}^n$. Let $\mathbf{x}^*$ be a point in D. Let U, the set of *active constraints* at $\mathbf{x}^*$, be

$$U = \{i : r_i(\mathbf{x}^*) = a_i\}.$$

A direction $\mathbf{d}$ is called a *KKT direction* at a point $\mathbf{x}^* \in D$ if $\nabla r_i(\mathbf{x}^*)^T\mathbf{d} \geq 0$ for all $i \in U$. Every direction feasible at $\mathbf{x}^*$ in the sense of the previous section is a KKT direction, although the converse does not always hold (see below). KKT stands for Karush-Kuhn-Tucker, the discoverers of the KKT conditions.

The conditions that we might expect would hold at a local minimum $\mathbf{x}^*$ are (1) that $\mathbf{x}^*$ be feasible, and (2) that $\nabla f(\mathbf{x}^*)\mathbf{d} \geq 0$ for all KKT directions $\mathbf{d}$. These two conditions are the *KKT conditions*. By a well-known principle of linear programming called Farkas's Lemma (see Papadimitriou and Steiglitz [1982]), the second condition is equivalent to requiring that there exist nonnegative numbers μ_i for each $i \in U$ such that

$$\nabla f(\mathbf{x}^*) = \sum_{i \in U} \mu_i \nabla r_i(\mathbf{x}^*).$$

We can define μ_i for $i \notin U$ by assuming $\mu_i = 0$ in this case. We can thus state the conditions as follows. Let $\mathbf{r}(\mathbf{x})$ denote $(r_1(\mathbf{x}), \ldots, r_s(\mathbf{x}))^T$; define $\mathbf{a} = (a_1, \ldots, a_s)^T$.

DEFINITION. *A point* $\mathbf{x}^*$ *is called a* **KKT point** *if the following conditions, called the* **KKT conditions**, *hold. Let* $\mathbf{p} = \mathbf{r}(\mathbf{x}^*) - \mathbf{a}$. *Then*

$$\mathbf{p} \geq \mathbf{0}$$

(i.e., $\mathbf{x}^*$ *is feasible) and there exist* $\mu_i \geq 0$ *for* $i = 1, \ldots, s$ *such that*

$$\mu_i p_i = 0 \quad \text{for } i = 1, \ldots, s$$

and

$$\nabla f(\mathbf{x}^*) = \mu_1 \nabla r_1(\mathbf{x}^*) + \cdots + \mu_s \nabla r_s(\mathbf{x}^*).$$

The variable μ_i *is called the* **multiplier** *for the ith constraint.*

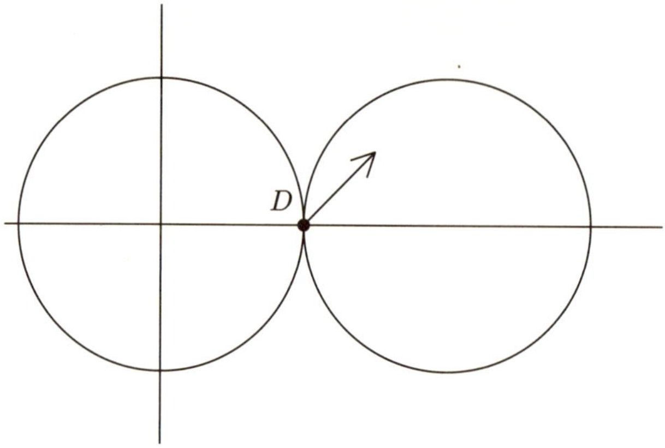

FIG. 1.3. KKT conditions can fail to be necessary.

This framework can be extended to cover equation constraints. Suppose that the domain is constrained by $t(\mathbf{x}) = a$, where t is a differentiable function. This equation constraint can be replaced by two inequalities, $t(\mathbf{x}) \geq a$ and $-t(\mathbf{x}) \geq -a$. Denote the multipliers for these two constraints μ_1, μ_2. Since both inequalities are satisfied as equations at a feasible point, the constraints on μ_1 and μ_2 are: $\mu_1, \mu_2 \geq 0$. The terms in the equation for $\nabla f(\mathbf{x}^*)$ are $\mu_1 \nabla t(\mathbf{x}^*) - \mu_2 \nabla t(\mathbf{x}^*)$. It then follows that we could replace $\mu_1 - \mu_2$ by a single multiplier μ for $\nabla t(\mathbf{x}^*)$ such that μ is not constrained in sign.

Thus, the above framework can be expanded to include equation constraints; each equation constraint has an unconstrained multiplier.

The KKT conditions are not in all cases necessary for local optimality, even if f and D are convex. A counterexample is as follows. Let the objective function on $\mathbb{R}^2$ be $f(x, y) = x + y$. Let the domain D be defined by the two constraints $x^2 + y^2 \leq 1$ and $(x-2)^2 + y^2 \leq 1$. This domain D is a singleton set, namely $\{(1,0)^T\}$. This point is clearly a local minimum. On the other hand, the objective function gradient, namely $(1,1)^T$, cannot be expressed as a linear combination of the gradient constraints. See Fig. 1.3. The KKT conditions break down in this case because there is a mismatch between feasible directions and KKT directions (see Exercise 1.3).

In the case that all the constraints are linear, the KKT directions and feasible directions are the same. In this case we can prove the necessity of the KKT conditions. There are also results giving conditions on nonlinearly constrained problems for which the KKT conditions are necessary—we refer the reader to Fletcher [1987]. We do not require the use of KKT conditions for nonlinear constraints except for one particular case discussed in Exercise 1.4 and again in Chapter 4. Here we state the main result for linear constraints.

THEOREM 1.4. *Suppose* $D \subset \mathbb{R}^n$ *is a nonempty convex domain specified by linear constraints of the form* $A\mathbf{x} \geq \mathbf{b}$. *Suppose* f *is a differentiable objective function and* $\mathbf{x}^*$ *is a local minimum. Then the KKT conditions must be satisfied at* $\mathbf{x}^*$.

PROOF. If $\mathbf{x}^*$ is feasible and the conditions cannot be satisfied, this implies that there is a KKT direction $\mathbf{d}$ such that for all $i \in U$, $\mathbf{a}_i^T\mathbf{d} \geq 0$ and $\nabla f(\mathbf{x}^*)^T\mathbf{d} < 0$. Here (as above) U is the set of indices i such that the ith coordinate of $A\mathbf{x} - \mathbf{b}$ is zero. The vector $\mathbf{a}_i$ denotes the ith column of A^T and is the gradient of the ith constraint. Then it is easy to see that $\mathbf{d}$ is a feasible direction in the sense of the previous section, because $\mathbf{a}_i^T(\mathbf{x}^* + \alpha\mathbf{d}) \geq b_i$ for all i and for all positive α small enough. By definition of gradient, $f(\mathbf{x}^* + \alpha\mathbf{d}) = f(\mathbf{x}^*) + \alpha\nabla f(\mathbf{x}^*)^T\mathbf{d} + o(\alpha)$ and the second term is negative. This shows that $\mathbf{x}^*$ cannot be a local minimum. ∎

EXAMPLES.

1. **Underdetermined least squares.** The problem of underdetermined least squares is the following: Minimize $f(\mathbf{x}) = \mathbf{x}^T\mathbf{x}$ subject to $A\mathbf{x} = \mathbf{b}$, where A is an $m \times n$ matrix with $m \leq n$. The constraints are linear so the preceding theorem applies. As noted above, the KKT framework handles equation constraints with multipliers unrestricted in sign. The gradient $\nabla f(\mathbf{x})$ evaluates to be $2\mathbf{x}$, and the gradient of the ith constraint is the ith column of A^T. Therefore, if $\mathbf{x}^*$ is a local minimum, there must exist a vector $\boldsymbol{\mu}$ such that
$$2\mathbf{x}^* = A^T\boldsymbol{\mu}.$$
Since $A\mathbf{x}^* = \mathbf{b}$, this implies that $AA^T\boldsymbol{\mu} = 2\mathbf{b}$. This latter equation has a unique solution provided that A is rank m. Once $\boldsymbol{\mu}$ is computed, then $\mathbf{x}^*$ is computed via $\mathbf{x}^* = A^T\boldsymbol{\mu}/2$. The vector $\mathbf{x}^*$ computed in this manner is in fact the global minimum—this follows from other results in this chapter because this optimization problem is convex.

2. **Linear programming.** Suppose that $f(\mathbf{x})$ is the linear function $\mathbf{c}^T\mathbf{x}$ and that the constraints are $A\mathbf{x} \geq \mathbf{b}$. Then the KKT conditions for this problem state that there exist multipliers $\mu_1, \ldots, \mu_s$ such that $\mathbf{c} = A^T\boldsymbol{\mu}$, $\boldsymbol{\mu} \geq \mathbf{0}$, and $\boldsymbol{\mu}^T(A\mathbf{x}^* - \mathbf{b}) = 0$. The vector $\boldsymbol{\mu}$ is called the vector of *dual variables*, and the preceding equations are called *complementary slackness* for linear programming.

The KKT conditions become especially valuable in the case of a convex domain with a convex objective function, as demonstrated by the following theorem. In the general case, the KKT conditions are not sufficient even for local minimality, but in the convex case, they are sufficient for *global* optimality.

THEOREM 1.5. *Let $D \subset \mathbb{R}^n$ be a convex domain specified by* (1.4) *in which each r_i differentiable, and $-r_i$ is convex. Let f be a convex differentiable objective function. Suppose $\mathbf{x}^* \in D$ is a KKT point. Then $\mathbf{x}^*$ is a global minimum.*

PROOF. Note that the requirement that $-r_i$ is convex implies that D is a convex domain; see Exercise 1.1. Choose any other point $\mathbf{x} \in D$; we want to show that $f(\mathbf{x}) \geq f(\mathbf{x}^*)$. Let $\mathbf{d} = \mathbf{x} - \mathbf{x}^*$; we claim that $\mathbf{d}$ is a KKT direction. In particular, choose an index i such that $r_i(\mathbf{x}^*) = a_i$.

Then $\mathbf{x}^* + \alpha\mathbf{d} \in D$ for all $\alpha \in [0, 1]$ by convexity, so $r_i(\mathbf{x}^* + \alpha\mathbf{d}) \geq a_i$, so $r_i(\mathbf{x}^* + \alpha\mathbf{d}) \geq r_i(\mathbf{x}^*)$. This means that the directional derivative of r_i in direction $\mathbf{d}$ must be nonnegative, i.e., $\nabla r_i(\mathbf{x}^*)^T\mathbf{d} \geq 0$. This shows that $\mathbf{d}$ is a KKT direction. Therefore, the KKT conditions assert that $\nabla f(\mathbf{x}^*)^T\mathbf{d} \geq 0$.

The inequality $\nabla f(\mathbf{x}^*)^T\mathbf{d} \geq 0$ implies the following fact by definition of derivative. For all $\epsilon > 0$ there exists an $\alpha > 0$ such that $f(\mathbf{x}^* + \alpha\mathbf{d}) \geq f(\mathbf{x}^*) - \epsilon\alpha$. Let $\alpha_i > 0$ be chosen so that

$$f(\mathbf{x}^* + \alpha_i\mathbf{d}) \geq f(\mathbf{x}^*) - \alpha_i/i$$

for $i = 1, 2, \ldots$ Now we can derive a sequence of inequalities for all i using (1.3) and the previous inequality.

$$\begin{aligned} f(\mathbf{x}^* + \alpha_i\mathbf{d}) &\leq (1 - \alpha_i)f(\mathbf{x}^*) + \alpha_i f(\mathbf{x}); \\ f(\mathbf{x}^*) - \alpha_i/i &\leq (1 - \alpha_i)f(\mathbf{x}^*) + \alpha_i f(\mathbf{x}); \\ \alpha_i f(\mathbf{x}^*) - \alpha_i/i &\leq \alpha_i f(\mathbf{x}); \\ f(\mathbf{x}^*) - 1/i &\leq f(\mathbf{x}). \end{aligned}$$

The preceding inequality holds for all $i = 1, 2, \ldots$ Therefore, $f(\mathbf{x}^*) \leq f(\mathbf{x})$ as desired. ∎

1.5. Quadratic programming and convexity

The theorems developed in the previous section for convex programming give powerful tools for algorithm development. For example, linear programming is a convex problem. In particular, let $\mathbf{c}$ be a vector in $\mathbb{R}^n$ and let f be the real-valued function $f(\mathbf{x}) = \mathbf{c}^T\mathbf{x}$. As mentioned in the last section, $f(\mathbf{x})$ is a convex function on $\mathbb{R}^n$—in fact, inequality (1.3) holds as an equation. Other facts mentioned in the last section imply that any region defined by linear constraints of the form $A\mathbf{x} \geq \mathbf{b}$ or $A\mathbf{x} = \mathbf{b}$ are convex regions. Using the results of the previous section, we arrive at the following well-known fact about linear programming: Any local minimum for a linear program is also a global minimum. Moreover, the complementary slackness conditions are sufficient to show global minimality.

It is very helpful to apply the methods of convexity to more general kinds of problems. In this section we give further characterizations of convexity and explain how they apply to quadratic programming.

Let f be a function from D to $\mathbb{R}$, where $D \subset \mathbb{R}^n$. Assume D is convex. Suppose f is twice-continuously differentiable (such a function is called a C^2 function). We denote the second derivative of f, called its *Hessian*, by f''. The Hessian $f''(\mathbf{x})$ is an $n \times n$ matrix whose entries are second partial derivatives. Because f is C^2, the Hessian is symmetric.

We define two special classes of square matrices. These definitions are most commonly applied to square symmetric matrices.

DEFINITION. *An $n \times n$ matrix H is said to be* **positive semidefinite** *(sometimes referred to as nonnegative definite) if for all $\mathbf{x} \in \mathbb{R}^n$, $\mathbf{x}^T H \mathbf{x} \geq 0$.*

DEFINITION. *An $n \times n$ matrix H is said to be* **positive definite** *if H is positive semidefinite, and the equation $\mathbf{x}^T H \mathbf{x} = 0$ implies that $\mathbf{x} = \mathbf{0}$.*

Some other useful definitions are as follows. Matrix H is *negative definite* if $-H$ is positive definite. The term *negative semidefinite* is defined in a similar fashion. A matrix is *indefinite* if it is none of the above. The term *indefinite* is sometimes also used if the matrix is not known to be any of the above, i.e., no definiteness condition is specified.

The importance of these definitions to optimization is contained in the following theorems.

THEOREM 1.6. *Let $f : D \to \mathbb{R}$ be a C^2 function, where D is a convex subset of $\mathbb{R}^n$. Suppose the Hessian $f''(\mathbf{x})$ is positive semidefinite for all $\mathbf{x} \in D$. Then f is a convex function.*

PROOF. Referring to (1.3), let $\lambda \in (0, 1)$, $\mathbf{x}, \mathbf{y} \in D$ be chosen. First, we establish the theorem for $n = 1$. In this case, x, y are real numbers, and without loss of generality, $x < y$. Let $z = (1 - \lambda)x + \lambda y$. The mean value theorem for differentiation tells us that there exist $z_1 \in (x, z)$ and $z_2 \in (z, y)$ such that $f(z) - f(x) = (z - x)f'(z_1)$ and $f(y) - f(z) = (y - z)f'(z_2)$. Since f'' is nonnegative (a 1-dimensional positive semidefinite matrix is a nonnegative number), $f'(z_2) \geq f'(z_1)$. This means that

$$\frac{f(y) - f(z)}{y - z} \geq \frac{f(z) - f(x)}{z - x}.$$

Simplifying this inequality yields exactly (1.3).

For the higher dimensional case, let

$$g(t) = f((1 - t)\mathbf{x} + t\mathbf{y}).$$

If we could establish that $g(t)$ is a convex function of one variable, then we could establish (1.3) for $f(\mathbf{x})$. To prove that $g(t)$ is convex, examine its second derivative, which is

$$g''(t) = (\mathbf{y} - \mathbf{x})^T f''((1 - t)\mathbf{x} + t\mathbf{y})(\mathbf{y} - \mathbf{x}).$$

The right hand side is always a nonnegative number since we are assuming the matrix in the middle is positive semidefinite. This establishes that g is convex, and hence so is f. ∎

In the special case that $f''(\mathbf{x})$ is positive definite for all $\mathbf{x}$, then we conclude that f is strictly convex because all the inequalities in the previous proof can be made into strict inequalities.

We now turn to the case of quadratic programming, that is, functions of the form $f(\mathbf{x}) = \frac{1}{2}\mathbf{x}^T H\mathbf{x} + \mathbf{c}^T\mathbf{x}$ where H is a symmetric matrix. It is easy to compute that $f''(\mathbf{x}) = H$ (independent of $\mathbf{x}$). This leads to the following theorem.

THEOREM 1.7. *Let H be an $n \times n$ symmetric matrix. The function $f(\mathbf{x}) = \frac{1}{2}\mathbf{x}^T H\mathbf{x} + \mathbf{c}^T\mathbf{x}$ is convex on $\mathbb{R}^n$ if and only if H is positive semidefinite.*

PROOF. First, assume H is positive semidefinite. Then the convexity of f follows immediately from the preceding theorem.

For the other direction, suppose that H is not positive semidefinite, i.e., there exists a vector $\mathbf{d}$ (necessarily nonzero) such that $\mathbf{d}^T H\mathbf{d} < 0$. Then if we set $\mathbf{x} = \mathbf{d}$ and $\mathbf{y} = -\mathbf{d}$ and $\lambda = 1/2$, then $0.5f(\mathbf{x}) + 0.5f(\mathbf{y})$ is seen to be $\frac{1}{2}\mathbf{d}^T H\mathbf{d}$, whereas $f(0.5\mathbf{x} + 0.5\mathbf{y}) = f(\mathbf{0}) = 0$, a violation of (1.3). ∎

If H is positive definite, then the function $f(\mathbf{x}) = \frac{1}{2}\mathbf{x}^T H\mathbf{x} + \mathbf{c}^T\mathbf{x}$ is strictly convex on $\mathbb{R}$.

In visualizing the relationship between positive definiteness an convexity, it is helpful to look at level curves. In Fig. 1.4, the level curves of the four quadratic functions $\frac{1}{2}\mathbf{x}^T H_i\mathbf{x}$ for $i = 1, 2, 3, 4$, where

$$H_1 = \begin{pmatrix} 4 & -0.4 \\ -0.4 & 2 \end{pmatrix}, \quad H_2 = \begin{pmatrix} -4 & 0.4 \\ 0.4 & -2 \end{pmatrix},$$

$$H_3 = \begin{pmatrix} 2 & -1.5 \\ -1.5 & -2 \end{pmatrix}, \quad H_4 = \begin{pmatrix} 4 & -0.4 \\ -0.4 & 0.04 \end{pmatrix}.$$

The plot shows level curves in the region $[-1, 1] \times [-1, 1]$. In these figures, an asterisk indicates the position in the square of the minimum objective function value. Notice that H_1 is positive definite, H_2 is negative definite, H_3 is indefinite, and H_4 is positive semidefinite but not positive definite. In H_4 there is a whole line of minima, so the asterisks are superimposed.

Exercises

1.1 Prove the facts listed earlier about convex functions and sets:

- The intersection of two convex sets is convex.

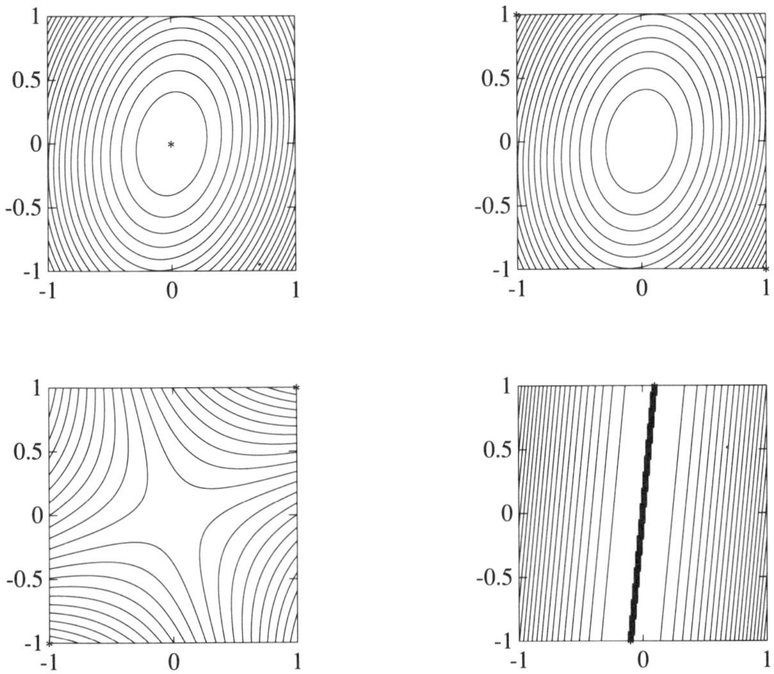

FIG. 1.4. Plots of four quadratic functions of two variables.

- The sum of two convex functions is convex.
- The pointwise maximum of two convex functions is convex.
- The function $f(\mathbf{x}) = \mathbf{a}^T\mathbf{x}$ is convex.
- If f is a convex function defined on a convex set E, then the set $D = \{\mathbf{x} \in E : f(\mathbf{x}) \leq a\}$ is convex for any choice of the real number a.

1.2 Let W be a nonempy convex compact subset of $\mathbb{R}^n$. Let $f : W \to \mathbb{R}$ be a continuous function. Let $\mathbf{a} \in \mathbb{R}^n$ be a fixed vector. Consider $p(\lambda)$ defined by the following *parameterized* optimization problem:

$$p(\lambda) = \min\{f(\mathbf{x}) : \mathbf{x} \in W; \mathbf{a}^T\mathbf{x} = \lambda\}.$$

Prove that $p(\lambda)$ is defined on a closed interval of $\mathbb{R}$ and is a continuous function on this interval.

1.3 Consider the optimization problem illustrated in Fig. 1.3: Minimize $f(x, y) = x + y$ subject to $x^2 + y^2 \leq 1$ and $(x-2)^2 + y^2 \leq 1$. What are the KKT directions at $(1, 0)$? What are the feasible directions? Do the KKT conditions hold at this point?

1.4 Verify that the KKT conditions must be satisfied at a local minimum of the following problem: Minimize $f(\mathbf{x})$ subject to $\mathbf{x}^T\mathbf{x} \leq 1$, where $f(\mathbf{x})$ is a differentiable real-valued function.

2

COMPLEXITY THEORY

The goal of complexity theory and algorithm analysis is to provide insight on how efficiently solutions to various problems like quadratic programming can be computed. In order to answer such questions, it is first necessary to have an understanding of a "problem" and an "efficient computation" for the problem. The term "complexity" refers to the amount of resources required by a computation and was introduced by Hartmanis and Stearns [1965]. In this book, running time is the only computational resource of interest.

This chapter gives a brief review of complexity theory. The books by Garey and Johnson [1979] and Papadimitriou and Steiglitz [1982] give a much more in-depth presentation of the material of this section. The complexity model assumed for Chapters 3 to 5 is the Turing machine model. The next few sections provide facts about Turing machine computation relevant for this book. In Section 2.4 we discuss other possible models of computation. Some of the definitions in this chapter are slightly nonstandard to simplify presentation, but are mathematically equivalent to the standard definitions.

2.1. Turing machines

We start with some definitions. A *problem* is a function $F : I \to B$, where I is the set of *instances* encoded as strings of characters. For example, linear programming is a "problem" in this sense. The set I would be quintuples of the form $(m, n, \mathbf{c}, A, \mathbf{b})$ encoded as strings of characters in some prespecified way. Each such quintuple is an "instance" of linear programming. Here, m, n are integers, $\mathbf{c}$ is a n-vector, A is an $m \times n$ matrix, and $\mathbf{b}$ is an m-vector. The value of $F(m, n, \mathbf{c}, A, \mathbf{b})$ is the minimum value of $\mathbf{c}^T\mathbf{x}$ over all choices of $\mathbf{x}$ satisfying $A\mathbf{x} \geq \mathbf{b}$. Thus, the set B in this example would be real numbers. The set B would also include an element to indicate that the input problem is infeasible, and another element to indicate that the problem is unbounded. We assume that elements of B are also encoded as strings of characters.

In the last paragraph, we left the precise meaning of instance undefined. From now on we require that the set I of instances for problems under consideration include *all possible* finite-length strings of characters over the alphabet for the problem. Here, the *alphabet* refers to the finite set of characters necessary to encode all useful instances. The alphabet for linear programming might include, for example, digits, a minus sign, a decimal point, commas and parentheses. When we say that instances of linear programming take the form $(m, n, \mathbf{c}, A, \mathbf{b})$, we mean that I includes all possible strings over the underlying alphabet, and we mean that F is supposed to map incorrectly formatted strings like "(4)3, −" to some default element of B that indicates incorrect formatting.

A *decision problem* is a problem (in the sense of the previous paragraphs) such that the output set B consists of two elements, **yes** and **no**. For example, consider the problem of determining whether a matrix is symmetric positive definite. The input set I would be $n \times n$ matrices. The function F would map the symmetric positive definite matrices to **yes** and all other matrices to **no**. A **yes**-*instance* of F is an instance that gets mapped to **yes**, and a **no**-*instance* is defined analogously. In the case of decision problems, it is customary to require that incorrectly-formatted input instances are **no**-instances.

The next step is to define the meaning of computing the function F. The standard model of computation is the *Turing machine*. A Turing machine's important features are its *tape* and its *finite-state controller*. The tape is a memory unit divided into tape cells. Each cell can hold one symbol chosen from a finite set of symbols (the finite set depends on the machine). The Turing machine has a *tape head* that can update one tape cell per computation step as directed by the finite-state controller. The controller has a finite list of *states*, $q_1, \ldots, q_s$. The controller has a finite list of rules telling it exactly what action should be taken based on the symbol currently under the tape head and on the current state. Possible actions include changing the contents of the current tape cell to some other symbol, moving the tape head left or right to a neighboring cell, changing the controller to a new state, halting, or some combination of these actions. The list of rules is meant to model the notion of a computer program. There must be exactly one rule that can be applied to any symbol-state combination (i.e., the action of the Turing machine is *deterministic*.) Fig. 2.1 gives a schematic picture.

A Turing machine is said to *compute* function $F : I \rightarrow B$ if, given an instance x of set I, it eventually produces $F(x)$ and halts. The machine finds x initially stored on its tape, one character per cell. The job of the machine is to leave $F(x)$ stored in memory, delimited in some fashion and encoded one character per tape cell.

2.2. P and NP

A *complexity class* is the class of problems (in the technical sense of the last section) that satisfy a certain resource bound. In this section we will

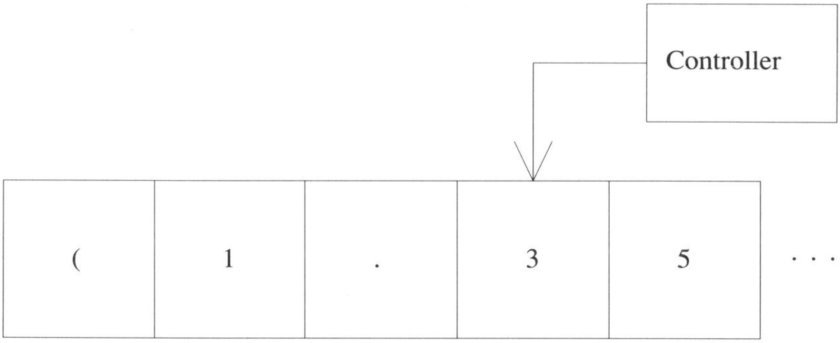

Fig. 2.1. A schematic of a Turing machine.

introduce the two complexity classes of use to this book, namely, **P** and **NP**.

DEFINITION. *The class* **P** *of problems is defined to be those decision problems* $F : I \to \{\textbf{yes}, \textbf{no}\}$ *such that a Turing machine* M *can compute* F*, and the number of steps required by* M *for this computation is bounded by* $p(n)$*, where* p *is a polynomial and* n *is the length of the input.*

In this definition we used the term *length of the input*. This is defined to be the number of characters in the string representation of the input instance. In general, if x is an object, then we use the notation $\mathrm{lth}(x)$ to denote the number of characters in the string encoding of the object.

EXAMPLE. Consider the problem mentioned above of determining if an $n \times n$ matrix A is symmetric positive definite. Checking for symmetry means verifying that $a_{ij} = a_{ji}$, which requires $O(n^2)$ numeric comparisons. Checking for positive definiteness may be done as follows. Factor A into LDL^T, where L is unit lower triangular and D is diagonal. (A *unit lower triangular* matrix $L = (l_{ij})$ is defined to have the property that $l_{ij} = 0$ if $i < j$ and $l_{ij} = 1$ if $i = j$. A *diagonal* matrix $D = (d_{ij})$ is defined to have the property that $d_{ij} = 0$ if $i \neq j$.)

If the factorization algorithm at any point computes a nonpositive diagonal entry of D, then A is not positive definite. Otherwise it is positive definite. The factorization requires $O(n^3)$ arithmetic operations (see Golub and Van Loan [1989] for more information). The number of data items in the input is n^2, and the number of arithmetic operations required is $O(n^3)$, so there appears to be a polynomial upper bound on the number of steps in terms of the length of the input.

Unfortunately, this argument does not demonstrate that the positive definiteness problem is in **P** for the following reason: Turing machines do not perform arithmetic operations on real numbers. A Turing machine operates on a finite set of symbols. We could imagine integer arithmetic being performed by a Turing machine using the standard elementary-school algorithms for long addition, long multiplication, etc. The number of steps to perform an arithmetic operation would therefore grow (either linearly or polynomially) with the number of digits in the operands. In this framework real numbers cannot be handled because they require an infinite number of digits.

Accordingly, we must assume that the real numbers handled by the Turing machine come from some finitely representable set. The easiest assumption to make is that our Turing machine for positive definiteness handles only rational numbers, and the numerators and denominators of the fractions are represented in binary, one bit per memory cell.

Now we return to the positive definiteness problem. We modify the statement of the original problem so that the input matrix A has rational entries. The comparisons necessary for the symmetry checking are straightforward—it is easy to test whether two rational numbers are equal. The LDL^T factorization requires only rational operations, so this can also be carried out with rational numbers. But now we face an additional difficulty: The rational numbers at an intermediate step of the factorization could require many more digits than the numbers in A. These longer numbers will adversely affect the running time analysis.

A naive analysis of the elimination algorithm would show that the number of digits could quadruple each elimination step, and therefore that the length of numbers could grow exponentially fast. In particular, the naive analysis is as follows. During the kth iteration of the factorization, we perform a computation like

$$a'_{ij} := a_{ij} - \frac{a_{ik}a_{kj}}{a_{kk}}$$

where the quantity on the left-hand side is a value computed on the kth iteration, and the values on the right-hand side were all computed on the $k-1$st iteration. Since an arithmetic operation involving two rational numbers in the worst case has an answer requiring as many digits as the total number of digits in all operands (for example, $59/77 + (97/53)(89/25) = 742916/102025$), we see that each entry of the matrix after updating with a computation like the one above could potentially have 4 times as many digits as the longest entry of the matrix before the kth iteration. Therefore, the number of the digits in matrix entries after k iterations grows like 4^k using this naive upper bound, which is faster than any polynomial.

Fortunately, there is an argument to show that the length of these intermediate numbers is bounded by a polynomial. We first state a theorem, originally due to Edmonds [1967] (the whole argument that follows is due to Edmonds).

THEOREM 2.1. *Let A be an $n \times n$ matrix with rational entries, and suppose that A requires at most ℓ binary digits to write down. Then the determinant of A requires no more than $O(n\ell)$ digits to write.*

PROOF. We compute the numerator and denominator of the determinant separately. Let d_i be the positive lowest common denominator among the entries of the ith row of A. Then d_i requires at most ℓ digits to write for each i (because it is the product of n integers whose total length is at most ℓ). Let A' be the integer matrix whose ith row is equal to d_i multiplied by the ith row of A. Then $\det(A) = \det(A')/(d_1 \cdots d_n)$, and $\det(A')$ is an integer. Thus, the remaining problem is to put an upper bound on the number of digits needed to represent $\det(A')$. This determinant is a sum of $n!$ terms. Each term is a product of n entries of A'. A product of n entries of A' has magnitude at most $d_1 \cdots d_n$ multiplied by the product of the same n entries of A. Thus, one term of the determinant has magnitude at most $2^\ell d_1 \cdots d_n$. Therefore, $\det(A')$ has magnitude at most $n!2^\ell d_1 \cdots d_n$.

Now we count the digits. The numerator in the expression for $\det(A)$ requires at most $n \lg n + \ell + n\ell$ digits. In this expression and for the rest of this book, $\lg x$ denotes the base-2 logarithm of x. The denominator requires at most $n\ell$ digits. This proves the theorem provided that $\ell \geq \lg n$ (a very natural requirement since it would be hard to write n^2 matrix entries with fewer than n^2 bits). ∎

Now, back to the LDL^T algorithm. It can be demonstrated that at each step of this algorithm, every quantity on hand is in fact the ratio of two *subdeterminants*, where a subdeterminant is defined to be the determinant of a square submatrix of A (see Exercise 2.1). Therefore, Edmond's theorem gives a bound on the size of intermediate numbers.

This bound applies only if the rational numbers are represented in lowest terms, because the previous theorem does not apply to an unreduced rational representation of a determinant. Therefore, it is necessary to compute greatest common divisors. This can be done in time polynomial in the number of digits using Euclid's algorithm; see Knuth [1973a].

If we assume that each arithmetic operation requires time proportional to the square of the number of digits, then each operation requires at most $O(n^2\ell^2)$ Turing machine steps. (Note that we are claiming that Turing machines can be programmed to carry out bit-by-bit arithmetic operations. This follows from the grade-school algorithms for these operations.) There are $O(n^3)$ operations total, so the whole algorithm requires $O(n^5\ell^2)$ Turing Machine steps. The length of the input, $\text{lth}(A)$, is at least $n^2 + \ell$ since there are n^2 matrix entries and ℓ bits of numeric data. Now it is a simple exercise to show that $O(n^5\ell^2)$ is bounded by a polynomial function of $n^2 + \ell$. This concludes the argument that testing positive definiteness is in **P**.

We have now exposed one of the key technical difficulties with regard to arguing about numerical optimization problems and the class **P**, namely, it is always necessary to examine the number of digits, and this may be a nontrivial task. Even for the simple problem of testing positive definiteness we required a fairly lengthy argument. See further remarks on this difficulty in Section 2.4.

A second problem with the definition of **P** is that optimization problems typically are not posed as decision problems. The standard way to turn an optimization problem into a decision problem is as follows. If the optimization problem is to minimize $f(\mathbf{x})$ subject to $\mathbf{x} \in D$ then the decision problem will be: Given f, D and a rational number ζ, does there exist an $\mathbf{x}^*$ in D such that $f(\mathbf{x}^*) \leq \zeta$?

We use the phrase *polynomial time* to describe an algorithm whose running time is a polynomial in the length of the input. In the theory of algorithms, an efficient algorithm is generally taken to mean a polynomial time algorithm. Above we saw an algorithm whose running time was $O(n^5\ell^2)$, which is polynomial but has a large exponent. For practical purposes it is of great interest to keep the degree of the polynomial small.

It has been found that when a problem has a polynomial time algorithm, it generally has an algorithm (not necessarily the same one!) that is efficient in practice.

Accordingly, showing that an optimization problem lies in **P** is an important goal and is considered a very positive result. The importance of **P** was first observed by Edmonds [1965] and Cobham [1965]. As we shall see in subsequent chapters, however, many optimization problems seem to lie outside **P**. It is therefore of interest to be able to determine for sure that a problem is not in **P**. There are no techniques known at present for proving that problems of interest to us are not in **P**. The best technique available is the theory of **NP**-completeness.

First, we define **NP**. Informally, a decision problem is said to be in **NP** if all of its **yes**-instances can be verified in polynomial time given a *certificate*. We use the following notation. If Λ is an alphabet, then Λ^* denotes the set of all possible finite-length strings that can be expressed with that alphabet. The definition of **NP** is as follows.

DEFINITION. *A decision problem* $F : I \to \{\mathbf{yes}, \mathbf{no}\}$ *is said to lie in* **NP** *if there exists a polynomial* $p(n)$*, a finite alphabet* Λ*, Turing machine* M *(the* **certificate checker***) running in polynomial time and computing a function* $G : I \times \Lambda^* \to \{\mathbf{yes}, \mathbf{no}\}$ *such that*

1. *For every* x *such that* $F(x) = \mathbf{yes}$ *there exists a string* $y \in \Lambda^*$ *(the* **certificate** *for* x*) such that the length of* y *is bounded by* $p(\mathrm{lth}(x))$ *and such that* $G(x, y) = \mathbf{yes}$*, and*

2. *For every* x *such that* $F(x) = \mathbf{no}$ *and for every string* $y \in \Lambda^*$*,* $G(x, y) = \mathbf{no}$.

REMARKS.

1. This definition says that every **yes**-instance has a certificate of polynomial length, and that there is a Turing machine that can check the certificates in polynomial time. Notice that no assumption is made about how the certificate is obtained.

2. For the Turing machine M that is supposed to compute a function of (x, y), we assume that x and y are written on the tape in order, one character per cell, and are separated by a delimiting symbol that does not occur in the alphabet for I nor in Λ.

3. In the definition we took the set Λ^* (the set of potential certificates) to be all possible strings over some alphabet. Normally, however, we think of the set of potential certificates as being a set $C \subset \Lambda^*$ of strings in a problem-dependent format. If G is given an input pair (x, y) such that y is in the wrong format, then $G(x, y) =$ **no**. For example, we might talk about a problem whose set of instances is the set of undirected graphs and whose set C of certificates is the set of lists of vertices of the graphs. What is meant technically by these statements is that I consists of all strings over some alphabet with a set of characters sufficient to represent undirected graphs (e.g., digits, parentheses, and commas). The set Λ might include digits and the comma. The set C might include only those strings involving digits and commas that do not start or end with commas or have two commas in a row. Finally, G will map any pair (x, y) with incorrect formatting in either entry to **no**.

4. Notice that **P** $\subset$ **NP** under these definitions. In particular, if the decision problem F is in **P**, then we can take Λ to be arbitrary. We define $G(x, y) = F(x)$ regardless of y. Turing machine M first deletes y from the tape and then simulates the Turing machine for F on x (i.e., its list of rules contains the rules for F as a subset, and it switches over to the rules for F after all the characters in y are deleted).

5. The set **P** was defined as decision problems, but it was easy to extend the notion of polynomial time to other problems. For the definition of **NP**, it seems to be crucial that only decision problems be considered.

6. The definition of **NP** treats **yes**-instances and **no**-instances asymmetrically. This means that, given a problem F in **NP**, if we change all of its **yes**-instances to **no**-instances and vice versa, the resulting problem is no longer necessarily in **NP**.

EXAMPLE. Consider the linear programming decision problem. The input instance is a sextuple $(m, n, \mathbf{c}, A, \mathbf{b}, \zeta)$, where $(m, n, \mathbf{c}, A, \mathbf{b})$ describe the linear program of minimizing $\mathbf{c}^T\mathbf{x}$ subject to $A\mathbf{x} \geq \mathbf{b}$, and A is $m \times n$. We assume that all these numbers are rational numbers encoded as binary fractions. The decision problem is to determine whether there exists a point $\mathbf{x}$ such that $A\mathbf{x} \geq \mathbf{b}$ and $\mathbf{c}^T\mathbf{x} \leq \zeta$.

We define the certificate to be a vector $\mathbf{x}$ with rational coordinates such that $A\mathbf{x} \geq \mathbf{b}$ and $\mathbf{c}^T\mathbf{x} \leq \zeta$ (note: we need not say how we come up with this vector $\mathbf{x}$). Now, it is clear that every **yes**-instance of the decision problem has such a certificate. Similarly, it is clear what the certificate-checking Turing machine must do, namely, plug the certificate $\mathbf{x}$ into the original LP and check whether $A\mathbf{x} \geq \mathbf{b}$ and $\mathbf{c}^T\mathbf{x} \leq \zeta$. If both inequalities hold, then the certificate-checker outputs **yes**, else it outputs **no**. Finally, we see that if $(m, n, \mathbf{c}, A, \mathbf{b}, \zeta)$ is a **no**-instance, then there cannot exist a vector $\mathbf{x}$ such that the certificate-checker will output **yes**. Thus, the certificate-checking Turing machine must be capable of standard arithmetic operations on rational numbers.

The only nontrivial part of this argument is to show that there exists a certificate $\mathbf{x}$ such that $\text{lth}(\mathbf{x})$ is bounded by a polynomial in $\text{lth}(m, n, \mathbf{c}, A, \mathbf{b}, \zeta)$ for a **yes**-instance. This proof follows from more general results about quadratic programming proved in Chapter 3 and Chapter 4, so we omit the proof for linear programming. The main idea of that proof is that at least one certificate will be the solution to a system of linear equations with coefficients drawn from $(\mathbf{c}, A, \mathbf{b}, \zeta)$. The solution to a system of linear equations can be computed with LDL^T factorization, and therefore can be expressed with rational entries whose numerators and denominators have a polynomial number of digits.

Most interesting combinatorial problems, when posed as decision problems, are in **NP**. This is because a problem is in **NP** if solutions can be verified efficiently (rather than computed). Verifying is usually simple for combinatorial problems.

2.3. NP-completeness

It is not known whether $\mathbf{P} = \mathbf{NP}$, but there is a theory that tells us which are the hardest problems in **NP**. If it turns out that $\mathbf{P} \neq \mathbf{NP}$ (as most researchers expect), then these problems would not be in **P** and would therefore require algorithms whose running time grows faster than any polynomial. A problem with no polynomially-bounded algorithm is generally called *intractable*.

We introduce the notion of transforming one problem to another.

DEFINITION. *A decision problem $F : I \to \{\mathbf{yes}, \mathbf{no}\}$ can be* **polynomially transformed** *into another decision problem $G : I' \to \{\mathbf{yes}, \mathbf{no}\}$ if there exists a function $f : I \to I'$ as follows. The function f can be computed by a polynomial-time Turing machine, and f has the property that*

$$F(x) = \mathbf{yes} \textit{ if and only if } G(f(x)) = \mathbf{yes}.$$

We remark that $\text{lth}(f(x))$ in this definition must be bounded by a polynomial in $\text{lth}(x)$ because f is computed by a polynomial time machine, i.e., a machine whose number of steps is bounded by a polynomial in $\text{lth}(x)$. Such a machine can produce at most one character of $f(x)$ per step.

As an example of a polynomial transformation, consider the *full-rank* problem. In this problem we are given an $m \times n$ rational matrix A with $m \geq n$, and we are asked to decide whether A has rank n. It is a standard result of linear algebra that A has rank n if and only if $A^T A$ is a symmetric positive definite matrix. Therefore, the full-rank problem can be polynomially transformed to the symmetric positive definiteness problem. We can express this transformation with the symbolism of the preceding definition. Set I would include encodings of $m \times n$ matrices with $m \geq n$, and set I' would include encodings of $n \times n$ matrices. Function F would map correctly-formatted encodings of matrices with full rank to **yes**, and function G would map correctly-formatted encodings of symmetric positive definite matrices to **yes**. Finally, function f would map a matrix A to $A^T A$. It can be easily argued that f can be computed by a polynomial-time Turing machine.

The decision problem G in the example of the last paragraph (symmetric positive definiteness) had been shown previously in this chapter to lie in **P**. In the discussion of **NP**-completeness below, we give examples of polynomial transformations in which the relevant decision problems are not known to be solvable in polynomial time. It is clear, however, that if F can be polynomially transformed to G, and G is in **P**, then so is F. For example, the polynomial transformation in the last paragraph proves that the full-rank problem lies in **P**.

We now introduce the CNF-satisfiability problem. A *boolean variable* x is a variable that takes as its value either 0 or 1. A *boolean formula* is built up from variables and from operations. The two binary operations are $\vee$ and $\wedge$ for OR and AND. The unary operation is $\neg$ for NOT. The operations have the usual interpretation, and the order of operations is determined by parentheses. An example of a formula is:

$$x \vee y \vee (z \wedge x \wedge \neg y).$$

A formula is said to be *satisfiable* if there is some *setting* (also called an *assignment*) of the variables (each variable set to 0 or 1) such that the formula evaluates to 1. The above formula is satisfiable: One satisfying assignment is $x = 1, y = 0, z = 0$.

A *CNF formula* is built up as follows. A *literal* is either a variable or its negation. A *clause* is an OR of literals. Finally, a CNF formula is an AND of clauses. Here is an example.

$$x \wedge (y \vee \neg x) \wedge (\neg x \vee \neg z).$$

This CNF formula has three clauses. The first clause has one literal, and the other two clauses each have two literals. The abbreviation CNF stands for conjunctive normal form.

The CNF-satisfiability decision problem, generally referred to as SAT, is the following: Given a CNF formula, determine whether it is satisfiable. This problem is clearly in **NP** because a **yes**-instance can be certified with the satisfying assignment. For example, the formula in the last paragraph is satisfiable, and a certificate is $x = 1, y = 1, z = 0$. The certificate-checking machine has to plug this satisfying assignment into the original formula and then evaluate the resulting boolean expression.

The following theorem is due to Cook [1971] and is often called Cook's Theorem.

THEOREM 2.2. *Every problem in* **NP** *can be polynomially transformed to SAT.*

We omit the proof, which is long and technical, and refer the reader to one of the standard texts cited at the beginning of this chapter. The idea of the proof is sketched out in the next few paragraphs. Let $F : I \to \{\textbf{yes}, \textbf{no}\}$ be a decision problem in **NP**. Let M be the certificate-checking machine specified in the definition of **NP**. Let $x \in I$ be an instance of F. The Turing machine T will transform x in polynomial time to an instance of SAT that "simulates" the action of M on (x, y) where y is a certificate for x. Since y is not known (indeed, y fails to exist if $F(x) = \textbf{no}$), the characters in y are represented by boolean variables. For example, the setting of boolean variable b_1 to 1 might stand for the fact that the first character of y is a left parenthesis. Many boolean variables are needed to capture the possible settings for each character of y, and T needs to produce CNF clauses that restrict the boolean variables only to consistent combinations. For example, if b_1 stands for the fact that the first character of y is a left parenthesis, and b_2 stands for the fact that the first character of y is '0', then these two variables cannot both be set to 1. Therefore, T must produce the clause $(\neg b_1 \vee \neg b_2)$ as part of its CNF formula.

Turing machine T must also produce clauses to ensure that M accepts (x, y). The computation of M cannot be simulated in advance since y is not known. Therefore, T represents the tape contents and states at each time step with more sequences of boolean variables. Clauses must be produced to ensure that the tape contents and state at time step i are the only valid successors from time step $i - 1$. Finally, T produces a clause to represent the fact that **yes** is left on the tape when M halts.

The goal is to end up with a CNF formula produced by T such that the formula has a satisfying assignment if and only if there exists a y as in the definition of **NP**. This formula, although long, is bounded by a polynomial in the length of x.

DEFINITION. *Suppose* $F : I \to \{\textbf{yes}, \textbf{no}\}$ *is a decision problem that lies in* **NP**, *and suppose that every problem in* **NP** *can be polynomially transformed to* F. *Then* F *is said to be* **NP-complete**.

Thus, Cook's theorem proves that SAT is **NP**-complete. Suppose that F is **NP**-complete, G is in **NP**, and F can be polynomially transformed to G. This implies that G is also **NP**-complete. The reason is that if E is some other problem in **NP**, then E can be transformed to F by assumption, and F can be transformed to G. Therefore E can be transformed to G by composing the two transformations (we omit the proof that the composition of two polynomial transformations is itself a polynomial transformation).

The conclusion is that in order to prove that a decision problem F is **NP**-complete, we must (1) show that F lies in **NP**, and (2) find a polynomial transformation from some **NP**-complete problem to F.

An **NP**-complete problem is the "hardest" problem in **NP** in the sense of polynomial transformations because of this definition. One would like to conclude that **NP**-complete problems are therefore intractable. This conclusion is unwarranted at present, because it is not known whether $\mathbf{P} = \mathbf{NP}$. If the answer turned out to be affirmative, then the **NP**-complete problems would also be in **P**, and therefore they would have efficient algorithms.

On the other hand, if it turns out that $\mathbf{P} \neq \mathbf{NP}$, all **NP**-complete problems would be intractable (Exercise: prove this statement). The conclusion $\mathbf{P} \neq \mathbf{NP}$ is generally believed to be more likely than the opposite conclusion. Accordingly, proving a problem to be **NP**-complete is about as far at present as one could go in terms of proving intractability.

We will now describe two **NP**-complete decision problems of use in later chapters. Both of these problems were originally proved **NP**-complete by Karp [1972]. Indeed, Karp's work showed researchers that many interesting problems are **NP**-complete, and that the class therefore deserves attention.

The CLIQUE problem is: Given an undirected graph G and an integer k, does G contain a set of k vertices that are all joined to one another by edges? Such a set of k vertices is called a clique.

THEOREM 2.3. *The CLIQUE problem is* **NP**-*complete.* ∎

PROOF. Our proof follows Karp's original proof. First, we must show that CLIQUE is in **NP**. Given a graph G and an integer k, the certificate will be the list of nodes of G that make up the clique. The Turing machine M (in the definition of **NP**), given an instance of clique and a certificate, verifies that every pair of vertices listed in the certificate is connected by an edge, and that there are k vertices listed in the certificate. It is easy to see that the certificate is polynomial in length (in fact, its length is shorter than the length of G), and that this certificate verification requires only polynomial time.

For the remainder of this book, when proving a problem is in **NP**, we describe its set of certificates but omit the description of what the certificate-checking Turing machine is supposed to do.

Next, we must explain how SAT is transformed to clique. Let q be an instance of SAT, that is, a boolean formula with m clauses $C_1 \wedge \cdots \wedge C_m$ and n variables, say $x_1, \ldots, x_n$. Assume that each variable occurs only once per clause (otherwise the formula can be simplified in a straightforward fashion). Suppose the total number of literals in $C_1, \ldots, C_m$ is t. We construct a graph G with t vertices as follows. Graph G has one vertex for each literal in each clause. We say that two literals are *consistent* if they are not opposites of

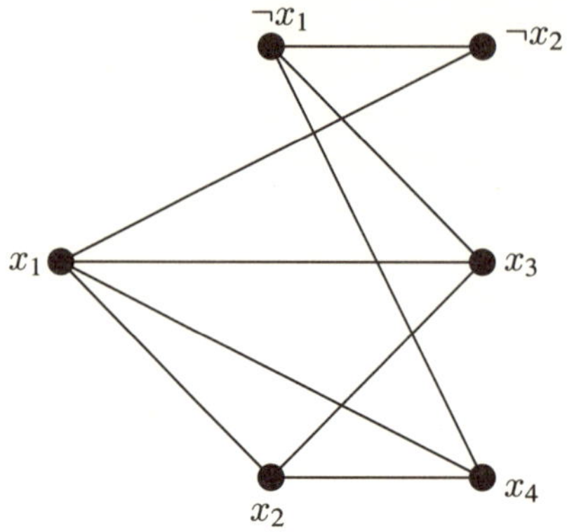

FIG. 2.2. The graph produced from $x_1 \wedge (\neg x_1 \vee x_2) \wedge (\neg x_2 \vee x_3 \vee x_4)$.

the same variable. In other words, x_i is consistent with every literal except $\neg x_i$. Then G has an edge between two vertices v_1, v_2 provided that (1) v_1, v_2 correspond to consistent literals, and (2) v_1, v_2 are literals in different clauses. An example of a graph produced by this transformation is illustrated in Fig. 2.2.

Finally, we have to produce the integer k that is part of a clique instance. We let $k = m$.

Graph G can be constructed from q by a polynomial-time Turing machine. Observe that G depends only on information gained from the structure of q (i.e., the way that the variables are arranged within the clauses). Accordingly, the transformation amounts to a series of pattern-matching operations on q to produce G. Such pattern-matching operations can be carried out by a Turing machine.

We now have to prove that q is a **yes**-instance iff G is. First, suppose G is a **yes**-instance. This means that G contains a clique of m vertices. Since the vertices corresponding to literals in the same clause are not connected, the clique must have exactly one literal from each clause. Since all these nodes are connected, these literals are all mutually consistent. This means that there is a way to set the variables so that all the literals in the clique are 1. For example, suppose v is a vertex in the clique corresponding to literal $\neg x_i$ occurring in clause C_j. If we set $x_i = 0$, this literal is true, and, moreover, the literal x_i (unnegated) does not occur anywhere in the clique. We have shown that the presence of the clique implies a setting of the variables that makes at least one literal in each clause equal to 1. Such a setting makes the formula q equal to 1 since the formula is CNF. Thus, the formula is satisfiable.

Conversely, suppose the formula is satisfiable. Choose some satisfying assignment; this assignment makes at least one literal in each clause equal to 1. For each clause, identify one vertex corresponding to a literal of that clause set to 1 by the satisfying assignment. It is easy to see that these vertices together form an m-clique.

Note that in order to compute the transformation of q to G we did not need to know a satisfying assignment for q nor a clique in G. This is important since the polynomial-time Turing machine computing the transformation does not have access to this information. The part of the proof in which we discussed cliques in G and satisfying assignments in q was the part in which we proved that the transformation has the property stated in the definition (i.e., that G has a k-clique iff q has a satisfying assignment). ∎

Another very useful problem is the SUBSET SUM problem, sometimes called the *0-1 knapsack* problem. This problem is: Given a list of nonnegative integers $a_1, \ldots, a_n$, and given an integer γ, is there a subset $J \subset \{1, \ldots, n\}$ such that

$$\sum_{j \in J} a_j = \gamma?$$

THEOREM 2.4. *SUBSET SUM is* **NP**-*complete.* ∎

PROOF. First, we must show that this problem is in **NP**. The certificate for a **yes**-instance is the subset J. It is easy to see what the certificate-checking machine must do.

Next, we prove that CLIQUE can be polynomially transformed to SUBSET SUM. Let (G, k) be an instance of CLIQUE, i.e., G is an undirected graph and k is an integer. Suppose G has n vertices and m edges. For convenience let the vertices of G be numbered $v_0, \ldots, v_{n-1}$. The list $\{a_0, \ldots, a_{t-1}\}$ will have $t = n + n^2$ entries. We assume that these integers are written out in binary. The digits of these numbers will be divided into $n+1$ *periods* of $n+2$ consecutive digits per period. The periods are numbered 0 to n so that $2^{i(n+2)+j}$ written in binary has a '1' in the ith period if j lies between 0 and $n+1$.

For $a_0, \ldots, a_{n-1}$, the nth (most significant) period will have a '1' in its least significant digit position, and zeros elsewhere. The remaining periods of these numbers are as follows. For a_i, a binary representation of the number n is inserted into the least significant portion of period i. The digit '1' is inserted into the least significant position of every other period j such that v_i is *not* adjacent to v_j.

Now we describe the remaining n^2 integers $a_n, \ldots, a_t$. In the remaining list, there are n occurrences of the number $2^{i(n+2)}$, as i runs between 0 and $n-1$. This number has a '1' in the least significant position of period i, and zeros elsewhere.

Finally, the right-hand side γ is taken to be the following integer. In the nth period, γ contains the binary representation of k (the requested clique size). In the other periods, γ contains a binary representation of n.

As in the previous proof, this transformation is carried out in polynomial time since it depends simply on structural properties. For the graph shown

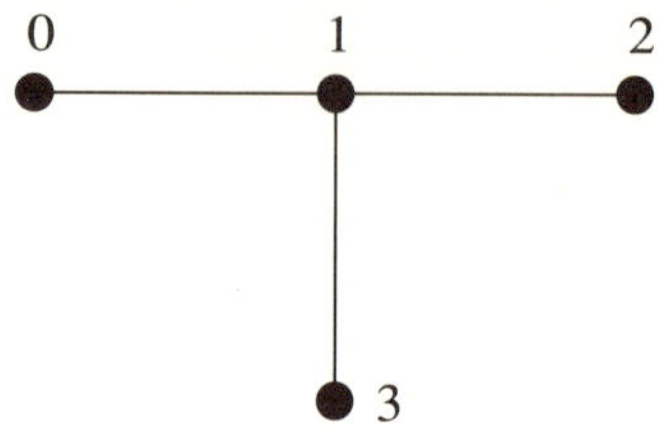

FIG. 2.3. An example graph for the subset sum transformation.

in Fig. 2.3 with $k = 3$, we get the following values for $a_0, \ldots, a_3$ and γ in binary:

$$\begin{aligned} a_0 &= 000001\,000001\,000001\,000000\,000100, \\ a_1 &= 000001\,000000\,000000\,000100\,000000, \\ a_2 &= 000001\,000001\,000100\,000000\,000001, \\ a_3 &= 000001\,000100\,000001\,000000\,000001, \\ \gamma &= 000011\,000100\,000100\,000100\,000100. \end{aligned}$$

The values for $a_4, \ldots, a_{11}$ in this example are omitted.

An overall comment about the construction is that addition of the a_i's cannot cause a period to carry over into the next period. The periods are long enough compared to the binary numbers stored in them that even if we added up all of the a_i's there would not be carry bits from one period to the next. This means that addition among the a_i's behaves like the addition of vectors with $n + 1$ components, one component for each period.

The claim is that G has a clique of size k iff list $\{a_0, \ldots, a_{t-1}\}$ has a subset J adding up to γ. First, suppose G has a clique of size k. We build J as follows. Take a_i as an element of J for each i such that v_i is in the clique. Observe that these numbers add up to k in the most significant period because there are k of them, and each has a 1 in this period. Notice that in the remaining periods, no sum exceeds n. Consider, for example, period j. If v_j is in the clique, then a_j has the number n in this period. Every other number among a_j for v_j in the clique has zeros in this period, because all of these other vertices are connected to v_j. The other case is that v_j is not in the clique. In this case, all of the numbers a_i for i selected have at most a '1' in this position, hence they add up to at most n.

Therefore, adding the numbers selected for J so far, we have k in the first period and numbers bounded by n in the remaining periods. Now we augment this to reach a sum of exactly γ by taking the appropriate number of entries from the list $\{a_n, \ldots, a_{t-1}\}$ to bring the sum in each period $0, \ldots, n-1$ up to n. This can be done because there are n separate copies of a '1' in each period in the remainder of the list.

Now we prove the other direction. Suppose a subset of list $\{a_0, \ldots, a_{t-1}\}$ adds up to γ. This list must have exactly k items of $\{a_0, \ldots, a_{n-1}\}$ to get agreement in the nth period. These numbers correspond to k vertices of G, which we claim correspond to vertices in a clique. Suppose not; suppose, for instance that a_0 and a_1 are selected for J, but there is no edge from v_0 to v_1 in G. This means that a_0 has a '1' in period 1, and a_1 has n in this period. Therefore, we get a sum of $n + 1$ in period 1. But now it is impossible to cancel out any part of this sum, so it is impossible for the whole list to add to γ. ∎

We now introduce additional terminology from complexity theory. A polynomial-time algorithm is said to be *strongly* polynomial time if the number of arithmetic operations of the algorithm is bounded by a polynomial in the dimensions of the matrices and vectors of the instance. This definition is not meaningful for nonnumeric problems. As an example, the algorithm for testing positive definiteness requires $O(n^3)$ arithmetic operations, so it would be strongly polynomial.

Suppose we had a problem $F : I \to B$, not necessarily a decision problem, and we are able to demonstrate that SAT could be *reduced* to F in the sense that SAT could be solved in polynomial time provided that F could. This would not prove that F is **NP**-complete since we might not be able to prove that F is in **NP**; indeed, F might not even be a decision problem. The term reserved for F in this situation is "**NP**-hard." Before giving the precise definition, we will prove a theorem to show the use of this definition.

The general quadratic programming problem is as follows: Given an $n \times n$ symmetric matrix H, an $m \times n$ matrix A, and vectors $\mathbf{b}, \mathbf{c}$, find the minimum of $\frac{1}{2}\mathbf{x}^T H\mathbf{x} + \mathbf{c}^T\mathbf{x}$ subject to $A\mathbf{x} \geq \mathbf{b}$. This can be stated as a "problem" in the technical sense given earlier. We have the following result concerning this problem.

THEOREM 2.5. *The general quadratic programming problem is* **NP**-*hard.*

PROOF. Let q be an instance of SAT, that is, a boolean formula with n variables $x_1, \ldots, x_n$ and m clauses. We produce a quadratic program $(H, \mathbf{c}, A, \mathbf{b})$ as follows. We denote the variables of the quadratic program by $y_1, \ldots, y_n$, one for each boolean variable. Linear constraints of the form $y_i \geq 0$ and $y_i \leq 1$ will make up the first $2n$ rows of A and $\mathbf{b}$. The goal is to force the y_i's to be either 0 or 1. For this purpose we define an objective function $f(\mathbf{y})$ by

$$f(y_1, \ldots, y_n) = y_1(1 - y_1) + \cdots + y_n(1 - y_n).$$

Notice that this formula is quadratic—indeed it is the same as $\frac{1}{2}\mathbf{y}^T H\mathbf{y} + \mathbf{c}^T\mathbf{y}$ with $H = \text{diag}(-2, -2, \ldots, -2)$ and $\mathbf{c} = (1, 1, \ldots, 1)^T$.

We have specified all of $(H, \mathbf{c})$ and the first $2n$ rows of $(A, \mathbf{b})$. Matrix A actually has $2n+m$ rows: the last m rows will contain one inequality for each clause of q. The inequality will be a sum of terms, one term per literal of the clause. If x_i occurs as a literal, then the corresponding term of the inequality will be y_i. If $\neg x_i$ occurs as a literal, then the term of the inequality will be

$(1 - y_i)$. The right hand side of the inequality will be 1. Thus, the clause $(x_1 \vee \neg x_2 \vee \neg x_3)$ will translate to the inequality $y_1 + (1 - y_2) + (1 - y_3) \geq 1$.

The claim is that the SAT formula is satisfiable if and only if the global minimum of the QP is 0. Suppose that the SAT formula is satisfiable, i.e., there is a setting of the variables such that each clause has at least one literal set to 1. We must show that the global minimum of the QP is 0. Set each variable y_i equal to the corresponding variable x_i (i.e., 0 or 1). For this setting, $f(\mathbf{y}) = 0$, the first $2n$ inequalities are clearly satisfied, and the last m inequalities are satisfied because at least one term on the left-hand side of each inequality is set to 1.

For the other direction, suppose that the global minimum of the QP is 0. Since the objective function is the sum of terms that are nonnegative on the domain, this implies that each term is zero, i.e., each y_i is 0 or 1. Then we claim that, by setting each x_i equal to the corresponding y_i, we have a satisfying assignment for the formula q.

This argument shows that SAT is reducible to QP. In particular, if we could solve QP in polynomial time, then we would also be able to solve SAT in polynomial time by transforming each instance of SAT to an instance of QP as described above. ∎

We wish to be more precise in our definition of **NP**-hard. Let $F : I \rightarrow B$ be a problem. We say that a Turing machine M can use F as an *oracle* if M is augmented with the following capability: M can compute $F(x)$ from x in a single step. Technically, M has a distinguished state o such that when M enters state o, the following occurs in one step. Machine M finds a string x written in memory starting in the cell under the tape head, and leaves state o with $F(x)$ written in place of x on the tape. We say that problem G is *Cook-reducible* to problem F (sometimes called polynomial-time Turing reducible) if G can be computed in polynomial time by a Turing machine that uses F as an oracle. Finally, a problem F is said to be **NP**-*hard* if SAT (or any other **NP**-complete problem) is Cook-reducible to F. Thus, the above proof shows that quadratic programming is **NP**-hard. The proof that QP is **NP**-hard is originally due to Sahni [1974].

A polynomial transformation defined earlier is a special case of a Cook-reduction (this is Exercise 2.4). Therefore, all **NP**-complete problems are **NP**-hard. The converse is not necessarily true.

In Chapter 4 we will show that a decision-problem version of quadratic programming is **NP**-complete. The reduction in the proof of Theorem 2.5 is the polynomial transformation of SAT to quadratic programming that we use to prove **NP**-completeness, and in Chapter 4 we address the question of membership in **NP**.

2.4. Models of computation

So far in this chapter we have spoken only of the Turing machine model. There are three reasonable criteria for selecting a model of computation:

1. Do results proved with the model correspond to practical experience?

2. Does the model correspond to actual computers?

3. Does the model correspond to the way that algorithm designers think about algorithms?

The Turing machine model has been successful with regard to criterion 1. In particular, the problems that we will prove to be in **P** (convex quadratic programming, l_2 trust region problems, etc.) are the same problems that practitioners solve routinely with efficient algorithms. There are still unsettling questions with regard to the Turing machine model. In particular, even though the model allows us to classify problems, it does not seem useful for classifying particular algorithms (e.g., the Turing machine model is not useful for distinguishing one interior point method from another).

The Turing machine is less successful with regard to the criterion of correspondence with actual computers. In particular, most modern computers carry out numerical algorithms via *floating-point* numbers, that is, numbers represented in scientific notation with fixed limits on the number of mantissa digits and on the range of the exponent. It should be observed that a Turing machine can simulate a "standard" digital computer and vice versa (provided the computer is equipped with enough memory for the simulation). These simulations are primarily of theoretical interest and do not lead to practical algorithms.

It might seem natural to create a model of computation based on a digital computer with floating-point numbers. Such models are in fact widely used to analyze errors in numerical analysis; see Golub and Van Loan [1989]. Such a model, however, would not be so useful for complexity analysis. The reason is that existing algorithms for many optimization problems require the precision of the arithmetic to depend on the size of the problem (see Chapter 3 for the main example of this phenomenon). If such an algorithm is used with standard floating point arithmetic on a large problem with many near-degeneracies, then the algorithm will probably fail to converge, execute a division by zero, or exhibit some other unwanted behavior. Such behavior must not be allowed if the aim is to construct a polynomial-time algorithm for a problem.

The Turing machine model fares extremely poorly with regard to the third criterion: No one has ever designed an optimization algorithm starting from a finite alphabet of tape symbols. In particular, most optimization algorithms are motivated by a series of mathematical formulas pertaining to real numbers.

The technical difficulty of squeezing real-number problems into a bit-by-bit format (illustrated by the LDL^T example of this chapter) has prompted some researchers to work in a model in which the primitive element of computation is a real number rather than a tape symbol. There are two principal real-number models occurring in the literature. In the first kind, part of the input is a black-box subroutine. For example, in studying optimization problems, we might say that f, the objective function, is an input to the minimization algorithm. The optimization algorithm may access f by passing $\mathbf{x}$ to a the subroutine and retrieving $f(\mathbf{x})$. This model of computation allows very general optimization problems to be considered. We return to this subject in Chapter 6.

The second kind of real number model assumes that the input is a finite list of real numbers. For example, linear programming would naturally be considered in such a model. Blum, Shub, and Smale [1988] redefine **P** and **NP** strictly in terms of real numbers. It is not yet clear that this kind of model can predict, for example, the sharp distinction between global minimization of convex QP and nonconvex QP. Thus, this model is not known yet to have the predictive power of the Turing machine model.

Another difficulty with real number models is that, at least in theory, there is no correspondence at all between these models and actual computers. This is because actual computers cannot carry the infinite number of digits of accuracy demanded by the models.

Exercises

2.1 (Golub and Van Loan [1989]) Let A be a symmetric nonsingular $n \times n$ matrix, and suppose $A = LDL^T$, where L is unit lower triangular and D is diagonal. Let d_{ii} denote the ith diagonal entry of D. Let A_i denote the upper left $i \times i$ submatrix of A. Show that $d_{ii} = \det(A_i)/\det(A_{i-1})$.

2.2 Prove that the following problem is **NP**-complete: Given a list of integers $a_1, \ldots, a_n$, is there a subset J such that

$$\sum_{j \in J} a_j = \sum_{j \notin J} a_j?$$

2.3 (Papadimitriou and Steiglitz [1982]) In the proof that the subset sum problem was **NP**-complete, it was necessary to use integers with many digits. Consider the special case of subset sum in which the integers $a_1, \ldots, a_n$ lie between 0 and k where k is a fixed constant. Show that the problem can be solved in time polynomial in n.

2.4 Suppose F and G are decision problems, and F can be polynomially transformed to G. Prove that F is Cook-reducible to G.

2.5 A *quadratic constraint* on a vector $\mathbf{x}$ is an inequality of the form $\frac{1}{2}\mathbf{x}^T B\mathbf{x} + \mathbf{c}^T\mathbf{x} \geq a$, where B is an $n \times n$ symmetric matrix, $\mathbf{c}$ is an n-vector, and a is a scalar. Show that the following decision problem is **NP**-hard. Given a domain D specified by a sequence of quadratic constraints, is D nonempty?

3

CONVEX QUADRATIC PROGRAMMING

At the end of Chapter 2, we proved that general quadratic programming is **NP**-hard, which is a negative result for algorithm designers. This result strongly suggests that there is no efficient algorithm for global optimization of a general quadratic program.

The traditional argument about why general quadratic programming appears to be intractable is as follows. Consider the quadratic program of minimizing

$$-\sum_{i=1}^{n} x_i^2$$

subject to constraints of the form $-1 \leq x_i \leq 1$ for all i. It can be verified that this problem has a local minimum at every vertex of the feasible region, i.e., at every point of the form $(\pm 1, \pm 1, \ldots, \pm 1)^T$. There are 2^n such points.

One might therefore expect that a general-purpose quadratic programming algorithm would require exponential time because it might have to loop through exponentially many local minima that are suboptimal. This argument is, of course, not a proof of the intractability of quadratic program: There might be an algorithm that could skip over all suboptimal local minima.

The preceding quadratic program and the quadratic program used in the proof of Theorem 2.5 are nonconvex, i.e., the matrices H in the objective functions are not positive semidefinite. In the convex case, the picture is much more positive. The phenomenon of exponentially many isolated local minima cannot occur in the convex case.

The goal of this chapter is to show that convex quadratic programming is in **P**. In the first section we study the Helgason-Kennington-Lall algorithm for the convex knapsack problem. This algorithm is strongly polynomial and uses the techniques developed in the last chapter in an interesting way. The analysis of its running time is straightforward. This algorithm is extremely efficient in practice.

In the second section, we turn to general convex quadratic programming. We show that an interior point algorithm for this problem solves it in polynomial time (but not in strongly polynomial time). The algorithm we propose has reasonable behavior in practice, but is not as efficient as a state-of-the-art implementation of recent interior point methods.

3.1. The HKL algorithm for convex knapsack

The *quadratic knapsack problem* (QKP) is the following:

$$\begin{aligned} &\text{minimize} && \tfrac{1}{2}\mathbf{x}^T D\mathbf{x} + \mathbf{c}^T\mathbf{x} \\ &\text{subject to} && \mathbf{a}^T\mathbf{x} = \gamma \text{ and} \\ &&& l_i \le x_i \le u_i \text{ for } i = 1, \ldots, n. \end{aligned} \tag{3.1}$$

In this problem, $\mathbf{x}$ is the vector of unknowns, D is an $n \times n$ diagonal matrix (i.e., all of the off-diagonal entries are 0), and $\mathbf{c}$, $\mathbf{a}$, $\mathbf{l} = (l_1, \ldots, l_n)^T$ and $\mathbf{u} = (u_1, \ldots, u_n)^T$ are n-vectors.

This problem is clearly a special case of quadratic programming. It is a convex problem if D is positive semidefinite. A diagonal matrix D is positive definite iff all of its diagonal entries are positive, and it is positive semidefinite iff all of its diagonal entries are nonnegative.

Helgason, Kennington and Lall [1980] discovered an $O(n \log n)$ algorithm for the positive definite case of QKP. In this section we will present a generalization of their algorithm for the positive semidefinite case. We refer to the algorithm as the HKL algorithm.

QKP arises in practical situations. One application is resource allocation, where γ represents the total amount of a resource, and each a_i is the consumption of this resource by the ith activity. Another application is as a subproblem for various other constrained optimization problems. See Cottle, Duvall and Zikan [1986] for an example and further references.

The first step toward solving a convex instance of (3.1) is to normalize the equation constraint. We assume without loss of generality that $a_i \neq 0$, because if $a_i = 0$ then variable x_i is completely decoupled from the rest of the problem and may be optimized independently. Since $a_i \neq 0$, we can scale x_i by $1/a_i$. Also, we observe that the objective function is *separable*, i.e., it may be written as a sum of functions of one variable:

$$\begin{aligned} &\text{minimize} && q_1(x_1) + \cdots + q_n(x_n) \\ &\text{subject to} && x_1 + \cdots + x_n = \gamma \text{ and} \\ &&& l_i \le x_i \le u_i \text{ for } i = 1, \ldots, n. \end{aligned} \tag{3.2}$$

Note that the scaling transformation does not alter the signs of the entries of D regardless of the signs of the a_i's. The functions $q_i(x)$ are therefore convex quadratic functions of one variable; we will assume that $q_i(x) = \frac{1}{2}d_i x^2 + c_i x$ ($d_i \ge 0$). Finally, we will assume that $l_i < u_i$ for each i, because if $l_i = u_i$ then the variable x_i is uniquely specified and could be eliminated from the problem. If $l_i > u_i$ then the problem is infeasible.

Since this is a convex problem, local minimality implies global minimality. Let us informally derive conditions for local minimality of a feasible point $\mathbf{x}^* = (x_1^*, \ldots, x_n^*)^T$. Suppose that $l_1 < x_1^* < u_1$ and $l_2 < x_2^* < u_2$.

Then it is easy to see that $q_1'(x_1^*)$ must be equal to $q_2'(x_2^*)$, otherwise we could increase one variable by ϵ and decrease the other by ϵ and improve the objective function. Such a perturbation of $\mathbf{x}^*$ would still be feasible.

Therefore, at a local optimum, there is some value λ such that for any x_i^* strictly between l_i and u_i, it must be the case that $q_i'(x_i^*) = \lambda$.

Now suppose that $x_1^* = l_1$ and $x_2^* \in (l_2, u_2)$. Using a similar argument about slight perturbations, we see that it is necessary that $q_1'(x_1^*) \geq q_2'(x_2^*)$. Finally, if $x_1^* = u_1$ and $x_2^* \in (l_2, u_2)$, then it is necessary that $q_1'(x_1^*) \leq q_2'(x_2^*)$.

The conclusion is that at a local minimum $\mathbf{x}^*$ there must exist a value λ such that

$$\begin{aligned} q_i'(x_i^*) &= \lambda \quad \text{if } l_i < x_i^* < u_i, \\ q_i'(x_i^*) &\geq \lambda \quad \text{if } x_i^* = l_i, \text{ and} \\ q_i'(x_i^*) &\leq \lambda \quad \text{if } x_i^* = u_i. \end{aligned} \tag{3.3}$$

These conditions together with the feasibility constraints are in fact equivalent to the KKT definitions; in particular, λ is the multiplier for the equation constraint $x_1 + \cdots + x_n = \gamma$. Therefore, feasibility and (3.3) are sufficient for global minimality.

These conditions suggest that λ plays a key role in solving the QKP problem. Suppose λ were known. Recall $q_i'(x) = d_i x + c_i$ with $d_i \geq 0$. Assume for a moment that $d_i > 0$. Then the equation $q_i'(x_i) = \lambda$ has a unique solution $\hat{x}_i = (\lambda - c_i)/d_i$. From $\hat{x}_i$, the value of x_i^* is uniquely determined by (3.3). In particular, it would be the case that $x_i^* = \hat{x}_i$ if $\hat{x}_i \in (l_i, u_i)$; $x_i^* = l_i$ if $\hat{x}_i \leq l_i$; and $x_i^* = u_i$ if $\hat{x}_i \geq u_i$.

Therefore, we can express the dependence of x_i^* on λ as function in the following simple form:

$$x_i^*(\lambda) = \text{median}\left(l_i, u_i, \frac{\lambda - c_i}{d_i}\right). \tag{3.4}$$

In the case that $d_i = 0$, x_i^* is not uniquely determined. Equation (3.3) tells us that if $\lambda < c_i$ then we must have $x_i^* = l_i$; if $\lambda > c_i$ then we must have $x_i^* = u_i$; and if $\lambda = c_i$ we can choose any $x_i^* \in [l_i, u_i]$. For this case we define a point-to-set function:

$$x_i^*(\lambda) = \begin{cases} \{l_i\} & \text{if } \lambda < c_i, \\ [l_i, u_i] & \text{if } \lambda = c_i, \text{ and,} \\ \{u_i\} & \text{if } \lambda > c_i. \end{cases} \tag{3.5}$$

Note that this function is an ordinary function at all but one point—at the point $\lambda = c_i$, its value is a whole interval.

We have now defined a vector-valued point-to-set function $\mathbf{x}^*(\lambda)$ so that for any choice of λ, this function satisfies (3.3). The point $\mathbf{x}^*(\lambda)$ is not, however, guaranteed to be feasible; in particular, the constraint that $x_1 + \ldots + x_n = \gamma$ might not be satisfied.

Define the point-to-set function

$$h(\lambda) = x_1^*(\lambda) + \cdots + x_n^*(\lambda).$$

Then we make the following important observation: $h(\lambda)$ is a nondecreasing function. An example of $h(\lambda)$ is plotted below in Fig. 3.1.

We should explain what these remarks mean for a point-to-set function. The meaning of the sum above is that $y \in h(\lambda)$ if there exist $y_1 \in x_1^*(\lambda), \ldots, y_n \in x_n^*(\lambda)$ such that $y = y_1 + \cdots + y_n$. The claim that h is nondecreasing means that if $\lambda_1 < \lambda_2$ and $y_1 \in h(\lambda_1), y_2 \in h(\lambda_2)$, then $y_1 \leq y_2$. To prove that $h(\lambda)$ is nondecreasing, we observe that each of the functions $x_i^*(\lambda)$ is nondecreasing—this follows from (3.4) and (3.5).

We can further describe $h(\lambda)$. We notice that each $x_i^*(\lambda)$ is a piecewise linear function with either one breakpoint (in the case of (3.5)) or two breakpoints (in the case of (3.4)). Therefore, $h(\lambda)$ is a piecewise linear nondecreasing point-to-set function. Finally, note that $h(\lambda) = u_1 + \cdots + u_n$ for λ large enough, and $h(\lambda) = l_1 + \cdots + l_n$ for λ small enough. Therefore we define

$$\begin{aligned} h(-\infty) &= l_1 + \cdots + l_n, \\ h(+\infty) &= u_1 + \cdots + u_n. \end{aligned}$$

Recall that $h(\lambda)$ expresses the left-hand side of the equation constraint of (3.2). Therefore, in order for this constraint to be satisfied, we must find λ^* such that $h(\lambda^*) = \gamma$ (more precisely, $\gamma \in h(\lambda^*)$). This is done via a binary search.

First, we sort all the breakpoints of each $x_i^*(\lambda)$: These are the points where λ is equal to $d_i l_i + c_i$ or equal to $d_i u_i + c_i$. There are at most $2n$ breakpoints, so sorting them requires $O(n \log n)$ steps (see Aho, Hopcroft and Ullman [1974] or Knuth [1973b]). This collection of breakpoints is also the set of breakpoints for h. Let t be the number of breakpoints; let their sorted values be $b_1, \ldots, b_t$. We observe that $h(b_t) = h(+\infty)$ (or perhaps $h(+\infty) \in h(b_t)$), because at the last breakpoint all of the x_i^* functions will be u_i. Similarly, $h(b_1) = h(-\infty)$ (or $h(-\infty) \in h(b_1)$).

Next, we do a binary search on the intervals. The goal is to find i such that $h(b_i) \leq \gamma \leq h(b_{i+1})$ (more precisely, there are $y \in h(b_i), y' \in h(b_{i+1})$ such that $y \leq \gamma \leq y'$).

The binary search requires two integers to act as brackets, which we call r and s. We initialize r to 1 and s to t. At each iteration we choose an integer midway between r and s, say m. Then we test whether $\gamma \in h(b_m)$, $\gamma < h(b_m)$, or $\gamma > h(b_m)$. In the case that $\gamma \in h(b_m)$, then we halt with $i = m$. If $\gamma < h(b_m)$, then we set $s = m$, else we set $r = m$ and repeat. This process continues until $r + 1 = s$. A total of $\lg t$ phases of this binary search are required.

The main computation at each phase of the binary search is the computation of $h(b_m)$. This value will either be a point or an interval, and it is easy to see from (3.4) and (3.5) how to evaluate $h(b_m)$. The number of operations to do this evaluation is bounded by $O(n)$.

This search will terminate with an i such that $h(b_i) \leq \gamma \leq h(b_{i+1})$, provided that initially

$$h(-\infty) \leq \gamma \leq h(+\infty). \tag{3.6}$$

If (3.6) fails to hold, then the original problem is infeasible. For example, $\gamma < h(-\infty)$ means that $\gamma < l_1 + \cdots + l_n$.

Accordingly, the binary search yields an interval $[b_i, b_{i+1}]$ such that λ^* lies in this interval (for a feasible problem). We want the exact value of λ^*. First, if $\gamma \in h(b_i)$ or $\gamma \in h(b_{i+1})$ then we set $\lambda^* = b_i$ or $\lambda^* = b_{i+1}$. Otherwise, we observe that h is a linear function between b_i and b_{i+1}. Moreover, the two coefficients of this linear function can be obtained from (3.4) and (3.5) in $O(n)$ operations. Then we can solve $h(\lambda^*) = \gamma$ to get λ^*.

Once we have λ^*, we use (3.4) and (3.5) to compute $\mathbf{x}^*$; in particular, $\mathbf{x}^* = (x_1^*(\lambda^*), \ldots, x_n^*(\lambda^*))^T$. A difficulty arises if $x_i^*(\lambda^*)$ is an interval for some i. In a degenerate case, $x_i^*(\lambda^*)$ could be multiply-valued for several indices i. For these values of i, $q_i(x) = c_i x$ with $c_i = \lambda^*$. For simplicity of notation, suppose this set of i's is $1, 2, \ldots, p$. Then x_i^* for $i = p+1, \ldots, n$ is determined already, and we are in the situation that

$$l_1 + \cdots + l_p + x_{p+1}^* + \cdots + x_n^* \leq \gamma$$

and

$$u_1 + \cdots + u_p + x_{p+1}^* + \cdots + x_n^* \geq \gamma$$

(because $\gamma \in h(\lambda^*)$). Accordingly, any setting of $x_1^*, \ldots, x_p^*$ such that $x_i^* \in [l_i, u_i]$ and such that $x_1^* + \cdots + x_n^* = \gamma$ is a global minimum. The algorithm for determining such a setting is left to the reader.

The main conclusion for this section is summarized with the following theorem.

THEOREM 3.1. *The HKL algorithm solves the convex quadratic knapsack problem in $O(n \log n)$ arithmetic steps and comparisons.*

PROOF. We have already argued that the HKL algorithm correctly solves the problem. The bound on the number of steps follows from totaling the running time for each part of the algorithm. Sorting the breakpoints requires $O(t \log t)$ steps, which is the same as $O(n \log n)$. There are $\lg t$ phases of binary search, each phase requiring $O(n)$ steps. Therefore, the binary search portion of the algorithm requires $O(n \log n)$ steps. Finally, once $[b_i, b_{i+1}]$ is discovered, an additional $O(n)$ steps are needed to compute λ^* and $\mathbf{x}^*$. ∎

REMARKS.

1. It was later discovered (Brucker [1984]) that this algorithm can actually be implemented in $O(n)$ steps. This has been left as an exercise (Exercise 3.3).

2. The HKL algorithm is an example of a strongly polynomial-time algorithm as described in the last chapter. See below.

EXAMPLE. We now work out an example of the HKL algorithm. Suppose the problem is to minimize $f(\mathbf{x}) = \frac{1}{2}\mathbf{x}^T D\mathbf{x} + \mathbf{c}^T\mathbf{x}$, where

$$D = \begin{bmatrix} 0 & 0 & 0 \\ 0 & 4 & 0 \\ 0 & 0 & 6 \end{bmatrix}$$

and $\mathbf{c} = (4, -5, -4)^T$. The inequality constraints are: $-1 \leq x_1 \leq 6$, $0 \leq x_2 \leq 8$, and $3 \leq x_3 \leq 7$. The equation constraint is $x_1 + x_2 + x_3 = 10$.

We first explicitly write out the x_i^* functions of λ.

$$x_1^*(\lambda) = \begin{cases} -1 & \text{if } \lambda < 4, \\ [-1, 6] & \text{if } \lambda = 4, \text{ and} \\ 6 & \text{if } \lambda > 4. \end{cases}$$

$$x_2^*(\lambda) = \text{median}\left(0, 8, \frac{\lambda + 5}{4}\right).$$

$$x_3^*(\lambda) = \text{median}\left(3, 7, \frac{\lambda + 4}{6}\right).$$

The list of breakpoints is $4, -5, 27, 14, 38$. These would be sorted. We will not explicitly provide the formulas h for this example, but it is depicted in Fig. 3.1. Then we would proceed with the binary search. We would find that $h(14) = 6 + (19/4) + 3 = 55/4$, which is larger than $\gamma = 10$. Next we would try $h(4) = [-1, 6] + (9/4) + 3 = [17/4, 45/4]$. This interval includes 10, so we have found λ^*, namely, $\lambda^* = 4$. We find that $x_2^*(4) = 9/4$ and and $x_3^*(4) = 3$, so x_1^* must be set to $10 - 3 - 9/4 = 19/4$. Thus, the global optimum is $(19/4, 9/4, 3)$.

Next, we wish to address the question of membership in **P**. The decision problem for the convex quadratic knapsack problem is as follows. Given an 8-tuple $(n, D, \mathbf{c}, \mathbf{a}, \mathbf{l}, \mathbf{u}, \gamma, \zeta)$ where n is an integer, all other entries are rational and D is positive semidefinite, is there an n-vector $\mathbf{x}$ such that $\mathbf{a}^T\mathbf{x} = \gamma$, $\mathbf{l} \leq \mathbf{x} \leq \mathbf{u}$, and $\frac{1}{2}\mathbf{x}^T D\mathbf{x} + \mathbf{c}^T\mathbf{x} \leq \zeta$? The HKL algorithm shows that the number of arithmetic operations and comparisons for this decision problem is bounded by a polynomial, so the only remaining hurdle is to get a bound on the size of the intermediate numbers of the algorithm.

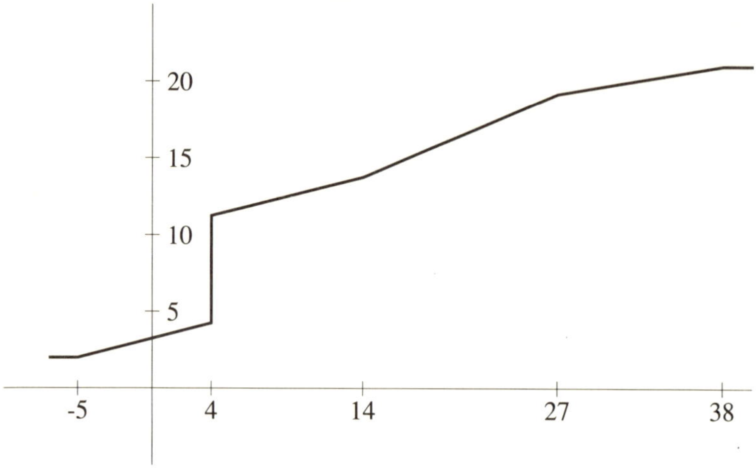

FIG. 3.1. The function $h(\lambda)$ for the example.

For the HKL algorithm a naive analysis on number growth suffices. The HKL algorithm, unlike LDL^T factorization, does not iterate arithmetic operations on intermediate results. This means that the worst-case analysis on the the growth of the numbers for each individual operation will demonstrate a polynomial bound. We conclude that the decision problem for the convex version of QKP is in **P**.

3.2. An interior point method

In the last section, we presented an algorithm for the convex version quadratic knapsack. In this section we present a polynomial-time interior point method for the general case of convex quadratic programming. Since linear programming is a special case of convex quadratic programming, the algorithm of this section also solves linear programming in polynomial time. It could also be used for the convex knapsack problem of last section, except that this use would be inefficient compared to the HKL algorithm. The phrase "interior point method" refers to an iterative algorithm in which each point $\mathbf{x}^{(k)}$ generated by the algorithm is in the interior of the feasible region.

The development of interior point methods was probably the most important breakthrough in optimization during the 1980's. The first interior point method for linear programming was discovered by Karmarkar [1984]. The method in this section is based on Newton's method, which was applied to linear programming first by Renegar [1986]. The algorithm in this section was proposed by Renegar and Shub [1988]. It is similar to an algorithm for linear programming proposed by Gonzaga [1989].

Other interior point algorithms for quadratic programming have been proposed by Kapoor and Vaidya [1986], Kojima, Mizuno and Yoshise [1989], Monteiro and Adler [1989], Ye and Tse [1989], and others.

The first polynomial-time algorithm for convex quadratic programming, the ellipsoid method, was discovered by Kozlov, Tarasov and Hačijan [1979]. The ellipsoid algorithm, as well as a discussion of its origins, is presented in Chapter 6.

The algorithm of this section is classified as a primal method with logarithmic barriers. It follows a central path and takes small steps.

Before turning to the interior point method, we discuss the format of the problem and some of its implications. We assume that the problem takes the form

$$\begin{aligned} \text{minimize} \quad & \tfrac{1}{2}\mathbf{x}^T H\mathbf{x} + \mathbf{c}^T\mathbf{x} \\ \text{subject to} \quad & A\mathbf{x} \geq \mathbf{b} \end{aligned} \tag{3.7}$$

where H is a positive semidefinite $n \times n$ matrix, and A is $m \times n$ matrix.

We assume that $(H, \mathbf{c}, A, \mathbf{b})$ are integers. In Chapter 2 we observed that it was necessary to make an assumption that the data is finitely represented. Up to now we have assumed that input data for numeric problems is rational. There is no loss of generality in assuming that it is integer, since we can simply clear the common denominators to change rational data to integer data. We assume integer data in this section because we can derive better bounds for this case; the bounds we derive are of the same order as the best bounds appearing in the literature.

We define a number L for an instance of convex QP as follows. We let L_0 be the number of bits required to write the integer-valued entries of $(H, \mathbf{c}, A, \mathbf{b})$. We let

$$L = L_0 + mn + n^2. \tag{3.8}$$

We assume that L is the length of the input in bits. This is because L_0 bits are required to write the numerical data, and $mn + n^2$ is the total number of entries in A and H (so we imagine that we need to separate the matrix entries in the input by "commas" and that there are $mn + n^2$ "commas").

Working with integer matrices automatically gives bounds on certain norms; the following lemma is an important example.

LEMMA 3.1. *Let W be an $n \times n$ nonsingular integer matrix and* $\mathbf{c}$ *an integer n-vector. Suppose that ℓ_0 bits are required to write down the entries of $(W, \mathbf{c})$. Then the solution* $\mathbf{x}$ *to* $W\mathbf{x} = \mathbf{c}$ *is a vector with rational entries. Moreover, these rational entries have a common denominator d less than $2^{\ell_0 + n \lg n}$ in magnitude, and the numerators with respect to d are each less than $2^{\ell_0 + n \lg n}$ in magnitude.*

PROOF. This follows from Cramer's rule for linear systems, which states that x_i is the quotient of two determinants a_i/d. The number d is the determinant of W and does not depend on i; a_i is the determinant of W with the ith column replaced by $\mathbf{c}$.

Since a_i and d are both integers, it suffices to prove an upper bound of $2^{\ell_0 + n \lg n}$ on their magnitude. A determinant is a sum of $n!$ terms, each of which is a product of n matrix entries. The product of n matrix entries is bounded by 2^{ℓ_0}. Therefore, the determinants have magnitude at most $n!2^{\ell_0}$. We do a calculation on this quantity:

$$\begin{aligned} n!2^{\ell_0} &< n^n 2^{\ell_0} \\ &\leq 2^{\ell_0 + n \lg n}. \end{aligned}$$

∎

We can make the preceding lemma more precise as follows. Suppose $W, \mathbf{c}, n, \ell_0$ are as above, and suppose that at most f rows of W have two or more nonzero entries (i.e., at least $n - f$ rows have exactly one nonzero entry). Then the magnitudes of the numerators and denominators are at most $2^{\ell_0 + f \lg f}$. This is because the determinant has at most $f!$ terms.

We now show how to apply this lemma to (3.7). For the purpose of the following lemma, we do not assume that (3.7) is convex.

LEMMA 3.2. *Suppose for a quadratic program of the form* (3.7) *that the problem is feasible and the global minimum is attained. Then there exists an optimal solution* $\mathbf{x}^*$ *with rational entries. Moreover, the size of the integers needed to express these rational numbers is at most* 2^{2L}. *This is true even when* H *is not positive semidefinite.*

PROOF. We have assumed that a global minimum exists. It is possible for there to be more than one global minimum. Among all global minima, let $\mathbf{x}^*$ be the global minimum such that the largest number of constraints among $A\mathbf{x} \geq \mathbf{b}$ are active. Here, *active* means that the constraint is satisfied as an equation. If there is more than one global minimum with the maximum number of active constraints, choose the minimum with the largest number of variables equal to zero. If there are still ties, they can be broken arbitrarily.

We claim that this particular $\mathbf{x}^*$ is the solution to a nonsingular system of linear equations. The remainder of the proof will expand on this claim. Suppose k constraints are active at $\mathbf{x}^*$ and suppose l entries of $\mathbf{x}^*$ are equal to zero. Then $\mathbf{x}^*$ satisfies the system of equations $M\mathbf{x}^* = \mathbf{b}'$, where M is the $(k + l) \times n$ matrix whose first k rows are equal to the active rows of A and whose last l rows express the constraints that certain coordinates are zero (i.e., these rows are rows of the identity matrix). Note that the entries of M and $\mathbf{b}'$ are either the original entries of A and $\mathbf{b}$ or 1's and 0's.

Let r be the rank of matrix M. Then the equations $M\mathbf{x} = \mathbf{b}'$ denote an $n - r$ dimensional affine subspace of $\mathbb{R}^n$. Call this subspace Y. Delete any dependent constraints from $M\mathbf{x} = \mathbf{b}'$ so that M has r rows.

If $r = n$ then Y contains the unique point $\mathbf{x}^*$. In this case we have established that $\mathbf{x}^*$ is the solution to a system of linear equations.

For the other case, consider the system

$$\begin{aligned} \text{minimize} \quad & f(\mathbf{x}) = \tfrac{1}{2}\mathbf{x}^T H\mathbf{x} + \mathbf{c}^T\mathbf{x} \\ \text{subject to} \quad & M\mathbf{x} = \mathbf{b}' \end{aligned} \tag{3.9}$$

We claim that system (3.9) has a unique global minimum, and this global minimum will be $\mathbf{x}^*$ (note that $\mathbf{x}^*$ is clearly feasible for this system). We also claim that if $M\mathbf{v} = \mathbf{0}$ and $\mathbf{v} \neq \mathbf{0}$ then $\mathbf{v}^T H\mathbf{v} > 0$.

For the geometric intuition behind this claim, examine Fig. 1.4. Let us consider a special case in which $M\mathbf{x} = \mathbf{b}'$ constrains $\mathbf{x}$ to lie in a two-dimensional affine space. Let us suppose, moreover, that the constraints $A\mathbf{x} \geq \mathbf{b}$, when restricted to this affine space, yield a set isomorphic to the square $[-1, 1]^2$. Then the second, third and fourth plots could not be examples of objective function f restricted to this space from the above construction. This is because there is a minimum on the boundary of $[-1, 1]^2$ in all of these plots, which is inconsistent with the requirement that $(M, \mathbf{b}')$ include all active constraints. Thus, only the first plot—in which the objective function is positive definite—is consistent with the way we have chosen $(M, \mathbf{b}')$.

We now give the algebraic proof of the claim. In this regard, consider a vector $\mathbf{x}^* + t\mathbf{v}$, where $M\mathbf{v} = \mathbf{0}$, $\mathbf{v} \neq \mathbf{0}$, $t \in \mathbb{R}$. We need to show that $f(\mathbf{x}^* + t\mathbf{v}) > f(\mathbf{x}^*)$ for all $t \neq 0$. We compute

$$f(\mathbf{x}^* + t\mathbf{v}) - f(\mathbf{x}^*) = \frac{1}{2}t^2\mathbf{v}^T H\mathbf{v} + (\mathbf{x}^{*T} H + \mathbf{c}^T)t\mathbf{v}. \tag{3.10}$$

First, observe that $(\mathbf{x}^{*T} H + \mathbf{c}^T)\mathbf{v} = 0$. Suppose not; suppose that this inner product is nonzero. Then the right-hand side of (3.10) is negative for all t small enough of the correctly-chosen sign, because the second term dominates the first when t is small. But $\mathbf{x}^* + t\mathbf{v}$ is feasible for (3.7) because $M\mathbf{v} = \mathbf{0}$ and all other constraints not included in M are inactive. This contradicts the global minimality of $\mathbf{x}^*$.

Therefore, we have established that the second term of (3.10) is zero for all $\mathbf{v}$ such that $M\mathbf{v} = \mathbf{0}$. Now, suppose that $\mathbf{v}^T H\mathbf{v} < 0$ for some such $\mathbf{v}$. Then the first term of the right-hand side of (3.10) would be negative for all choices of t, again contradicting global minimality.

Therefore, we have now shown that $\mathbf{v}^T H\mathbf{v} \geq 0$ whenever $\mathbf{v} \neq \mathbf{0}$ and $M\mathbf{v} = \mathbf{0}$. In addition, we have shown that $\mathbf{v}^T H\mathbf{x}^* + \mathbf{c}^T\mathbf{v} = 0$ for the same set choices of $\mathbf{v}$. These two facts prove that $\mathbf{x}^*$ is a global minimum for (3.9).

Now we want to show that $\mathbf{x}^*$ is in fact the unique global minimum of (3.9). Suppose, for example, that $f(\mathbf{x}^* + \mathbf{v}) = f(\mathbf{x}^*)$ for a vector $\mathbf{v} \neq \mathbf{0}$, $M\mathbf{v} = \mathbf{0}$. Using the preceding arguments, we can conclude that $\mathbf{v}^T H\mathbf{v} = 0$ and $\mathbf{v}^T H\mathbf{x}^* + \mathbf{c}^T\mathbf{v} = 0$. Therefore, $f(\mathbf{x}^* + t\mathbf{v}) = f(\mathbf{x}^*)$ for all choices of t. Notice that $\mathbf{x}^* + t\mathbf{v}$ has at least as many active constraints as $\mathbf{x}^*$ and at least as many coordinates equal to zero because $M\mathbf{v} = \mathbf{0}$. This means that for the original problem (3.7) we can replace $\mathbf{x}^*$ with $\mathbf{x}^* + t\mathbf{v}$, where t is chosen so that $\mathbf{x}^* + t\mathbf{v}$ is feasible and has one more active constraint or one more coordinate equal to zero.

The existence of this point contradicts the choice of $\mathbf{x}^*$, which was supposed to be the global minimum with the largest number of active constraints and largest number of zero coordinates. Therefore, we conclude that $\mathbf{x}^*$ is the unique global minimum of (3.9). Moreover, we conclude that $\mathbf{v}^T H\mathbf{v} > 0$ whenever $\mathbf{v} \neq \mathbf{0}$, $M\mathbf{v} = \mathbf{0}$.

Since $\mathbf{x}^*$ is a global minimum of (3.9) it must satisfy the KKT conditions for this problem, which are: There exists an r-vector $\boldsymbol{\mu}^*$ such that

$$\begin{aligned} H\mathbf{x}^* + \mathbf{c} &= M^T\boldsymbol{\mu}^*, \\ M\mathbf{x}^* &= \mathbf{b}'. \end{aligned}$$

We claim that the coefficient matrix for this system, namely

$$W = \begin{pmatrix} H & -M^T \\ M & 0 \end{pmatrix}$$

is nonsingular. In particular, suppose that the system

$$\begin{aligned} H\mathbf{x} - M^T\boldsymbol{\mu} &= \mathbf{0}, \\ M\mathbf{x} &= \mathbf{0} \end{aligned}$$

has a nontrivial solution. Clearly if $\mathbf{x} = \mathbf{0}$ then $\boldsymbol{\mu} = \mathbf{0}$ since M^T is rank r. Thus, a nontrivial solution has $\mathbf{x} \neq \mathbf{0}$; taking the inner product of the first equation with $\mathbf{x}$ shows that $\mathbf{x}^T H\mathbf{x} - \mathbf{x}^T M^T\boldsymbol{\mu} = 0$; the second equation implies that the second term drops out, so $\mathbf{x}^T H\mathbf{x} = 0$ for a nonzero vector $\mathbf{x}$ such that $M\mathbf{x} = \mathbf{0}$. Earlier we observed that there cannot exist a vector with these properties.

Thus, $\mathbf{x}^*$ is the solution to a nonsingular system of linear equations. The dimension of this system is at most $2n$ since M has at most n rows. The entries can be written down in $2L_0$ bits, because the entries of M require as many bits as the entries of A. Thus, by the previous lemma, the maximum absolute value of any entry in the solution to this system is less than $2^{2L_0+(2n)\lg(2n)}$ which is bounded by 2^{2L}. ∎

The remainder of this section is organized as follows.

1. In 3.2.1, the main loop of the Renegar-Shub algorithm is described. This is the computational heart of the interior point method. The main loop produces a series of iterates $(\mathbf{x}^{(k)}, t^{(k)})$, $k = 0, 1, 2, \ldots$ such that $\mathbf{x}^{(k)}$ converges to a global minimum of (3.7) and $t^{(k)}$ converges to 0.

2. In 3.2.2, we describe how to find a global minimum $\mathbf{x}^*$ from $\mathbf{x}^{(k)}$ when k is large enough. In a nondegenerate case, the global

minimum is obtained by solving a system of linear equations arising from the KKT conditions.

3. In 3.2.3 the initialization procedure for the algorithm, that is, the selection of $(\mathbf{x}^{(0)}, t^{(0)})$, is described.

4. In 3.2.4, the number of iterations and work per iteration is analyzed.

5. In 3.2.5 we argue that the entire algorithm can be carried out with numbers that have a polynomial number of digits.

6. In 3.2.6, we explain conjectures and related results concerning the existence of strongly polynomial-time algorithms for (3.7).

The material in 3.2.2 to 3.2.5 is more of theoretical interest than practical interest for the following reason. The various bounds derived in those sections are good enough that we can claim polynomial time for the algorithm, but they are too large to be practical. The interior point methods that work well in practice often do not have polynomial time bounds, a fact that indicates that our understanding of these algorithms is far from complete.

3.2.1. *The main loop of the Renegar-Shub algorithm*

We return to the case that H in (3.7) is positive semidefinite. Let Π denote the feasible region of (3.7), that is, the set of points satisfying $A\mathbf{x} \geq \mathbf{b}$. Let $\operatorname{Int}\Pi$ denote the interior of Π, that is, the set of points satisfying $A\mathbf{x} > \mathbf{b}$. We have already assumed that the data $(H, \mathbf{c}, A, \mathbf{b})$ in this problem is integer. We make the following additional assumptions.

ASSUMPTION 3.1. *Matrix A has rank n.*

ASSUMPTION 3.2. *The feasible set Π is compact.*

ASSUMPTION 3.3. *The set* $\operatorname{Int}\Pi$ *is nonempty.*

Later on we show that these assumptions may be made without loss of generality. There is one more assumption to make, which we postpone until more definitions are presented.

We define a *potential function*

$$h(\mathbf{x}, t) = \frac{1}{2}\mathbf{x}^T H\mathbf{x} + \mathbf{c}^T\mathbf{x} - t\sum_{1=1}^{m} \ln(\mathbf{a}_i^T\mathbf{x} - b_i)$$

for points in $\operatorname{Int}\Pi$. Here, $\mathbf{a}_i$ is the ith column of A^T. Notice that for any fixed $t > 0$, this function tends to $+\infty$ as the boundary of Π is approached. We would like to establish a few properties of the function.

We remark that the following three lemmas figure into the development of the algorithm only insofar as they define $\boldsymbol{\xi}(t)$. The remaining information provided by these lemmas is presented to give the reader background on the construction of the algorithm.

LEMMA 3.3. *For any fixed choice of* $t > 0$, *the function* h *is strictly convex on* $\operatorname{Int}\Pi$.

PROOF. We compute the Hessian of h; it is equal to

$$H + t\sum_{i=1}^{m} \frac{\mathbf{a}_i\mathbf{a}_i^T}{(\mathbf{a}_i^T\mathbf{x} - b_i)^2}.$$

This is the sum of positive semidefinite matrices. Since the vectors $\mathbf{a}_i$ span $\mathbb{R}^n$, it is actually positive definite. Then the result follows from Theorem 1.6. ∎

LEMMA 3.4. *For any fixed choice of* $t > 0$, *the function* h *has a unique minimum.*

PROOF. Since h is strictly convex, the minimum is unique if it exists (by Theorem 1.3). Therefore we only have to prove existence. Let $\mathbf{x}^{(0)}$ be a strictly interior feasible point, and consider the set $C = \{\mathbf{x} \in \operatorname{Int}\Pi : h(\mathbf{x}, t) \le h(\mathbf{x}^{(0)}, t)\}$. This is a closed nonempty set. Moreover, since Π is bounded, this set is compact. Therefore, the minimum of h exists. ∎

Let $\boldsymbol{\xi}(t)$ be the minimum of $h(\mathbf{x}, t)$ for a fixed t. Notice that as $t \to 0$, there must be an accumulation point of $\boldsymbol{\xi}(t)$ by compactness. This point must be a global minimum. Convergence of $\boldsymbol{\xi}(t)$ as $t \to 0$ is the subject of Exercise 3.4.

LEMMA 3.5. *Let* $\hat{\mathbf{x}}$ *be an accumulation point of* $\boldsymbol{\xi}(t)$ *as* $t \to 0$. *Then* $\hat{\mathbf{x}}$ *is a global minimum for the original problem* (3.7).

PROOF. Let $t^{(1)}, t^{(2)}, \ldots$ be a sequence of positive parameter values tending to zero such that $\boldsymbol{\xi}(t^{(k)})$ converges to $\hat{\mathbf{x}}$. Let U be the subset of indices indicating *active* constraints at $\hat{\mathbf{x}}$, i.e., $\mathbf{a}_i^T\hat{\mathbf{x}} = b_i$. At $\boldsymbol{\xi}(t)$ the gradient of $h(\mathbf{x}, t)$ is zero (gradient with respect to $\mathbf{x}$), i.e.,

$$H\boldsymbol{\xi}(t) + \mathbf{c} - t\sum_{i=1}^{m} \frac{\mathbf{a}_i}{\mathbf{a}_i^T\boldsymbol{\xi}(t) - b_i} = \mathbf{0}.$$

This is because no constraints are active since the point $\boldsymbol{\xi}(t)$ is interior. Write $\mathbb{R}^n$ as a direct sum $S \oplus T$ where S is the span of $\{\mathbf{a}_i : i \in U\}$ and T is the orthogonal complement of S. Let π_T denote orthogonal projection of $\mathbb{R}^n$ onto T; then

$$\pi_T(H\boldsymbol{\xi}(t) + \mathbf{c}) = t\sum_{i \notin U} \frac{\pi_T(\mathbf{a}_i)}{\mathbf{a}_i^T\boldsymbol{\xi}(t) - b_i}.$$

Note that this sum is only taken over $i \notin U$ since $\pi_T(\mathbf{a}_i) = \mathbf{0}$ for $i \in U$. The denominators of the above fraction are bounded below by a positive number for small $t^{(k)}$ because $\boldsymbol{\xi}(t^{(k)})$ is close to $\hat{\mathbf{x}}$ and these constraints are inactive. Accordingly, the above sum tends to zero as $t^{(k)}$ approaches 0. We conclude that $H\hat{\mathbf{x}}+\mathbf{c}$ lies in S. Now, let π_S denote projection onto S, and we compute

$$\pi_S(H\boldsymbol{\xi}(t^{(k)})+\mathbf{c}) = t^{(k)} \sum_{i=1}^{m} \frac{\pi_S(\mathbf{a}_i)}{\mathbf{a}_i^T \boldsymbol{\xi}(t^{(k)}) - b_i}. \tag{3.11}$$

Notice that this sequence comes arbitrarily close to the cone

$$C = \left\{ \sum_{i \in U} \mu_i \mathbf{a}_i : \mu_i \geq 0 \right\}$$

because the coefficients for $\mathbf{a}_i$ in the right-hand side of (3.11) for $i \in U$ are all nonnegative, and the coefficients for $i \notin U$ tend to zero.

Since C is closed and the sequence $H\boldsymbol{\xi}(t^{(k)})+\mathbf{c}$ converges, it converges to a point in C. Thus, $H\hat{\mathbf{x}}+\mathbf{c}$ lies in C. This statement, combined with the feasibility of $\hat{\mathbf{x}}$, is exactly equivalent to the KKT conditions, hence $\hat{\mathbf{x}}$ is a global minimum. ∎

We use a *parameterized Newton's method* to solve the system of equations $\nabla h(\mathbf{x}, t) = \mathbf{0}$, where ∇ indicates differentiation with respect to the vector $\mathbf{x}$. In general, *Newton's method* for solving $f(\mathbf{x}) = \mathbf{0}$ is an iterative method in which the iteration update is

$$\mathbf{x}^{(k+1)} = \mathbf{x}^{(k)} - f'(\mathbf{x}^{(k)})^{-1} f(\mathbf{x}^{(k)}).$$

Here, $f'(\mathbf{x}^{(k)})$ is the $n \times n$ matrix called the Jacobian of f. The superscript '-1' in the previous formula indicates matrix inversion.

Suppose we want to solve $\chi(\mathbf{x}, t) = \mathbf{0}$ as t tends to zero, where χ is a differentiable function from a subset of $\mathbb{R}^{n+1}$ to $\mathbb{R}^n$. Then we choose a sequence $t^{(k)}$ converging to zero and use the following iteration

$$\mathbf{x}^{(k+1)} = \mathbf{x}^{(k)} - \chi'(\mathbf{x}^{(k)}, t^{(k+1)})^{-1} \chi(\mathbf{x}^{(k)}, t^{(k+1)}) \tag{3.12}$$

where differentiation is with respect to $\mathbf{x}$. This is the parameterization of Newton's method that we apply to $\nabla h(\mathbf{x}^{(k)}, t)$. In this case, χ' in the above formula is the Hessian of h. We set $t^{(k)} = t^{(0)}\alpha^k$ for a parameter $0 < \alpha < 1$ to be chosen below.

The idea of the parameterized Newton's method is as follows. We know that tracing the path $\boldsymbol{\xi}(t)$ as $t \to 0$ will lead to a global minimum. Instead of computing $\boldsymbol{\xi}(t)$ exactly, we compute a sequence $\mathbf{x}^{(k)}$ of approximations to this vector as $t \to 0$. See Fig. 3.2 for an example. In this figure, there are two unknowns ($n = 2$). The asterisks indicate the iterates $\mathbf{x}^{(k)}$ generated by the algorithm. The dotted line indicates $\boldsymbol{\xi}(t)$. The constraints are indicated by the bounding line segments. This plot was generated from the Renegar-Shub algorithm, except that α, $\mathbf{x}^{(0)}$ and $t^{(0)}$ were chosen differently from the discussion of this section to make the figure clearer.

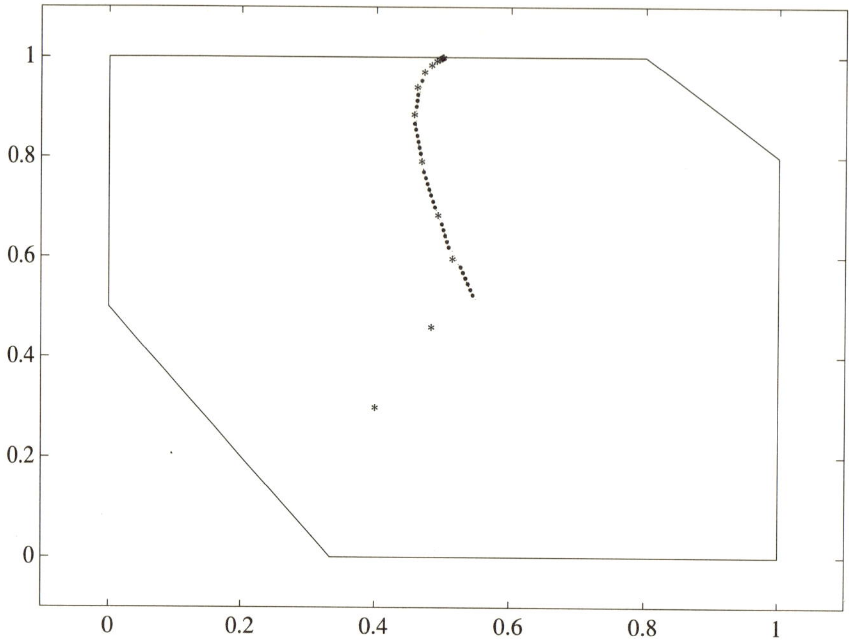

FIG. 3.2. A diagram indicating convergence of the main loop.

For the parameterized Newton method, the derivative of $\nabla h(\mathbf{x}, t)$ is needed; this matrix was calculated to be

$$H + t\sum_{i=1}^{m} \frac{\mathbf{a}_i\mathbf{a}_i^T}{(\mathbf{a}_i^T\mathbf{x} - b_i)^2},$$

which is symmetric positive definite and hence invertible. Let $\Delta(\mathbf{x})$ denote the diagonal matrix whose ith diagonal entry is $\mathbf{a}_i^T\mathbf{x} - b_i$. Then we can write this derivative as $H + tA^T\Delta(\mathbf{x})^{-2}A$.

Define $\nu(\mathbf{x}, t)$ by

$$\nu(\mathbf{x}, t) = \|\Delta(\boldsymbol{\xi}(t))^{-1}A\mathbf{x}\|_2.$$

Note that for a fixed t, this function is a norm on $\mathbb{R}^n$. The norm puts the most weight on directions affecting the near-active constraints at $\boldsymbol{\xi}(t)$.

We say that a point $\mathbf{x}$ is a *good approximation* to $\boldsymbol{\xi}(t)$ if $\nu(\mathbf{x} - \boldsymbol{\xi}(t), t) \leq 1/20$. Suppose we have a good approximation $\mathbf{x}$ to $\boldsymbol{\xi}(t)$ for some particular choice of t. We argue that a good approximation $\mathbf{x}$ for $\boldsymbol{\xi}(t)$ must be interior to the feasible region. From the relation

$$\|\Delta(\boldsymbol{\xi}(t))^{-1}A(\mathbf{x} - \boldsymbol{\xi}(t))\|_2 \leq 1/20,$$

we see that each component of $\Delta(\boldsymbol{\xi}(t))^{-1}A(\mathbf{x} - \boldsymbol{\xi}(t))$ is no more than $1/20$, i.e.,

$$\left| \frac{\mathbf{a}_i^T(\mathbf{x} - \boldsymbol{\xi}(t))}{\mathbf{a}_i^T\boldsymbol{\xi}(t) - b_i} \right| \leq 1/20.$$

The numerator of the above fraction is equal to $(\mathbf{a}_i^T\mathbf{x} - b_i) - (\mathbf{a}_i^T\boldsymbol{\xi}(t) - b_i)$. Changing the absolute value to a chain of inequalities yields

$$\frac{19}{20}(\mathbf{a}_i^T\boldsymbol{\xi}(t) - b_i) \leq \mathbf{a}_i^T\mathbf{x} - b_i \leq \frac{21}{20}(\mathbf{a}_i^T\boldsymbol{\xi}(t) - b_i). \tag{3.13}$$

The first inequality implies that $\mathbf{a}_i^T\mathbf{x} - b_i > 0$ so $\mathbf{x} \in \operatorname{Int}\Pi$.

We now make our final assumption about the initial conditions for the main loop of the algorithm.

ASSUMPTION 3.4. *We have a point* $\mathbf{x}^{(0)}$ *that is a good approximation to* $\boldsymbol{\xi}(t^{(0)})$.

The idea is to decrease t on each iteration and generate a sequence of good approximations to $\boldsymbol{\xi}(t)$.

We now focus on one particular iteration of the parameterized Newton method. We drop the superscript k. We use the superscript '+' to denote values at the next iteration. We assume that we have a good approximation for $\boldsymbol{\xi}(t)$. The goal for the next few pages is to demonstrate that a step of the parameterized Newton method will yield a good approximation for $\boldsymbol{\xi}(t^+)$.

As mentioned earlier, we have chosen $t^+ = \alpha t$. Let $\chi(\mathbf{x}, t^+)$ denote $\nabla h(\mathbf{x}, t^+)$; this can be written as

$$\chi(\mathbf{x}, t^+) = H\mathbf{x} + \mathbf{c} - t^+ A^T \Delta(\mathbf{x})^{-1}\mathbf{e}$$

where $\mathbf{e}$ is the m-vector with all entries equal to 1. The parameterized Newton step is

$$\mathbf{x}^+ = \mathbf{x} - \chi'(\mathbf{x}, t^+)^{-1}\chi(\mathbf{x}, t^+). \tag{3.14}$$

Computing (3.14) is the main operation of the algorithm. Following Renegar and Shub, we claim $\mathbf{x}^+$ is a good approximation for $\boldsymbol{\xi}(t^+)$. We state this as a theorem.

THEOREM 3.2. *Suppose* $\mathbf{x}$ *is a good approximation to* $\boldsymbol{\xi}(t)$. *Suppose* $\mathbf{x}^+$ *is generated from* $\mathbf{x}$ *via* (3.14). *Then, for the right choice of* α *(see* (3.16)*),* $\mathbf{x}^+$ *will be a good approximation to* $\boldsymbol{\xi}(t^+)$.

PROOF. This proof will last several pages and will include several supporting results in the process. The proof relies on a general result concerning the convergence of Newton's method due to Renegar and Shub, who derive it from work by Smale [1986]. Their theorem is not proved here.

In order to state the theorem, we need some background on the notation. Let $L(A, B)$ be the space of linear functions from A to B, where A and B are two vector spaces over the real numbers. For example $L(\mathbb{R}^p, \mathbb{R}^q)$ is identified with the set of $q \times p$ matrices. Let $L^p(A, B)$ denote the set of p-linear functions from A to B, that is, the set of functions from A^p to B that are linear in each element. Let $f : \mathbb{R}^n \to \mathbb{R}^n$ be an infinitely-differentiable function. It is well known that f' is a function $\mathbb{R}^n \to L(\mathbb{R}^n, \mathbb{R}^n)$, i.e., the function f' carries a vector to an $n \times n$ matrix. One way to interpret the pth derivative of f, denoted by $D^p f$, is as a map from $\mathbb{R}^n$ to $L^p(\mathbb{R}^n, \mathbb{R}^n)$. The derivative map $D^p f(\mathbf{x})$ is computed as follows. The value of this map applied to vectors $(\mathbf{v}_1, \ldots, \mathbf{v}_p)$ in $(\mathbb{R}^n)^p$ is

$$\sum_{\pi} \frac{\partial}{\partial x_{\pi(1)}} \cdots \frac{\partial}{\partial x_{\pi(p)}} f(\mathbf{x}) v_{1,\pi(1)} \cdots v_{p,\pi(p)} \tag{3.15}$$

where the sum is taken over all maps $\pi : \{1, \ldots, j\} \to \{1, \ldots, n\}$.

In general, if $\phi : (\mathbb{R}^n)^p \to \mathbb{R}^n$ is a multilinear map and ν is a norm on $\mathbb{R}^n$, then we can define the induced norm $\bar{\nu}(\phi)$ by

$$\bar{\nu}(\phi) = \sup\{\nu(\phi(\mathbf{v}_1, \ldots, \mathbf{v}_p)) : \nu(\mathbf{v}_i) \leq 1\}.$$

THEOREM 3.3 (RENEGAR AND SHUB) *Let $f : U \to \mathbb{R}^n$ be an analytic function, where U is an open subset of $\mathbb{R}^n$. Let ν be a norm on $\mathbb{R}^n$. Let $\mathbf{x}$ be a vector in U. Let*

$$\beta = \nu(Df(\mathbf{x})^{-1} f(\mathbf{x}))$$

and

$$\gamma = \sup_{p \geq 2} \bar{\nu}\left(\frac{1}{p!} Df(\mathbf{x})^{-1} D^p f(\mathbf{x})\right)^{1/(p-1)}$$

Suppose $\delta > 0$ satisfies $\beta \leq \delta/2 \leq 1/(40\gamma)$, and suppose that the ν-norm ball centered at $\mathbf{x}$ of radius 4δ lies in U. Let $\mathbf{x}^{(0)}$ be any point in the ν-norm ball centered at $\mathbf{x}$ of radius δ. Then the Newton iteration starting at $\mathbf{x}^{(0)}$ converges to a root $\mathbf{x}^*$ that satisfies $\nu(\mathbf{x}^* - \mathbf{x}) \leq 3\beta/2$, and the convergence rate is quadratic:

$$\nu(\mathbf{x}^{(k)} - \mathbf{x}^*) \leq 3.5(1/2)^{(2^k)}\delta. \blacksquare$$

The remarkable feature of this theorem (compared to other results about Newton's method) is that it gives convergence results for Newton's method based on norm estimates at a single point $\mathbf{x}$, and moreover, $\mathbf{x}$ can be different from the Newton starting point and from the root of f.

We now compute β and γ for the problem at hand. The results will be stated as lemmas so that we can reuse them below. The center of the ball of Theorem 3.3 is taken to be the point $\boldsymbol{\xi}(t)$. The norm we use is $\nu(\cdot, t)$. The function f is taken to be $\chi(\cdot, t^+)$, i.e., the gradient $\nabla h(\cdot, t^+)$. Recall that $f'(\boldsymbol{\xi}) = H + t^+ A^T \Delta(\boldsymbol{\xi})^{-2} A$, a symmetric positive definite matrix.

LEMMA 3.6. *(Calculation of β.) Let t_1, t_2 be positive numbers and $\boldsymbol{\xi} \in \operatorname{Int}\Pi$ a vector. Suppose f is defined by*

$$f(\mathbf{x}) = H\mathbf{x} + \mathbf{c} - t_1 A^T \Delta(\mathbf{x})^{-1}\mathbf{e}.$$

Suppose $\boldsymbol{\xi}$ satisfies

$$H\boldsymbol{\xi} + \mathbf{c} - t_2 A^T \Delta(\boldsymbol{\xi})^{-1}\mathbf{e} = \mathbf{0}.$$

Let $\nu(\mathbf{x})$ be the norm

$$\nu(\mathbf{x}) = \|\Delta(\boldsymbol{\xi})^{-1}A\mathbf{x}\|_2.$$

Then

$$\nu(Df(\boldsymbol{\xi})^{-1}f(\boldsymbol{\xi})) \le \left|\frac{t_1 - t_2}{t_1}\right| \cdot \sqrt{m}.$$

PROOF. We have:

$$f'(\boldsymbol{\xi})^{-1}f(\boldsymbol{\xi}) = (H + t_1 A^T \Delta(\boldsymbol{\xi})^{-2}A)^{-1}(H\boldsymbol{\xi} + \mathbf{c} - t_1 A^T \Delta(\boldsymbol{\xi})^{-1}\mathbf{e}).$$

Let M denote the matrix $\Delta(\boldsymbol{\xi})^{-1}A$. Then

$$f'(\boldsymbol{\xi})^{-1}f(\boldsymbol{\xi}) = (H + t_1 M^T M)^{-1}(H\boldsymbol{\xi} + \mathbf{c} - t_1 M^T\mathbf{e}).$$

We know that $H\boldsymbol{\xi} + \mathbf{c} - t_2 M^T\mathbf{e} = \mathbf{0}$. Therefore, we can simplify the last parenthesized expression:

$$f'(\boldsymbol{\xi})^{-1}f(\boldsymbol{\xi}) = (H + t_1 M^T M)^{-1}((t_2 - t_1)M^T\mathbf{e}).$$

Finally,

$$\begin{aligned}\nu(f'(\boldsymbol{\xi})^{-1}f(\boldsymbol{\xi})) &= |t_1 - t_2| \cdot \|M(H + t_1 M^T M)^{-1}M^T\mathbf{e}\|_2 \\ &\le |t_1 - t_2| \cdot \|M(H + t_1 M^T M)^{-1}M^T\|_2 \cdot \|\mathbf{e}\|_2.\end{aligned}$$

Clearly $\|\mathbf{e}\|_2 = \sqrt{m}$. Let us examine matrix $M(H + t_1 M^T M)^{-1}M^T$. We can write a singular value decomposition $M = U\Sigma V^T$. Here, U is an $m \times m$ orthogonal matrix, V is an $n \times n$ orthogonal matrix and Σ is an $m \times n$ diagonal matrix. Recall that M is rank n (because A is rank n), hence all the diagonal entries of Σ are nonzero. Let $\bar{\Sigma}$ be the first n rows of Σ; this is an invertible diagonal matrix, and $\Sigma^T\Sigma = \bar{\Sigma}^2$. Then

$$\begin{aligned}\|M(H + t_1 M^T M)^{-1}M^T\|_2 &= \|U\Sigma V^T(H + t_1 V\Sigma^T\Sigma V^T)^{-1}V\Sigma^T U\|_2 \\ &= \|U\Sigma(V^T HV + t_1\bar{\Sigma}^2)^{-1}\Sigma^T U\|_2 \\ &= \|\Sigma(V^T HV + t_1\bar{\Sigma}^2)^{-1}\Sigma^T\|_2 \\ &= \|\bar{\Sigma}(V^T HV + t_1\bar{\Sigma}^2)^{-1}\bar{\Sigma}\|_2 \\ &= \|(\bar{\Sigma}^{-1}V^T HV\bar{\Sigma}^{-1} + t_1 I)^{-1}\|_2.\end{aligned}$$

The matrix in parentheses is symmetric positive definite, and all of its eigenvalues are greater than or equal to t_1 because $V^T HV$ is positive semidefinite. Therefore, the above 2-norm is bounded above by $1/t_1$. We conclude that

$$\nu(f'(\boldsymbol{\xi})^{-1}f(\boldsymbol{\xi})) \leq \frac{|t_1 - t_2| \cdot \sqrt{m}}{t_1}.$$

∎

We apply this lemma immediately to our problem. In the lemma, we take $t_1 = t^+$. We take $t_2 = t$; notice that $\boldsymbol{\xi}(t)$ satisfies the condition of the lemma by definition. We take f in the lemma to be $\chi(\cdot, t^+)$. Therefore, we find that β of Theorem 3.3 is bounded by $\sqrt{m} \cdot |t^+ - t|/t^+$.

Next we compute γ, also via a lemma.

LEMMA 3.7. *(Computation of γ) Let $\boldsymbol{\xi}$ be a vector in* Int Π. *Let f be as in* Lemma 3.6. *Let $p \geq 2$ be an integer. Let ν be the norm*

$$\nu(\mathbf{x}) = \|\Delta(\boldsymbol{\xi})^{-1}A\mathbf{x}\|_2.$$

Then

$$\bar{\nu}\left(\frac{1}{p!}f'(\boldsymbol{\xi})^{-1}D^p f(\boldsymbol{\xi})\right) \leq 1.$$

PROOF. For the function f under consideration, the terms $H\mathbf{x} + \mathbf{c}$ drop out when we are looking at second and higher derivatives. Therefore, we need to compute the higher derivatives of $-t_1 A^T \Delta(\mathbf{x})^{-1}\mathbf{e}$. We use (3.15) to discover that

$$D^p f(\boldsymbol{\xi})(\mathbf{v}_1, \ldots, \mathbf{v}_p) = (-1)^{p+1} t_1 p! \sum_{i=1}^{m} \frac{\mathbf{a}_i}{(\mathbf{a}_i^T \boldsymbol{\xi} - b_i)^{p+1}} [(\mathbf{a}_i^T \mathbf{v}_1) \cdots (\mathbf{a}_i^T \mathbf{v}_p)].$$

We can rewrite this as

$$D^p f(\boldsymbol{\xi})(\mathbf{v}_1, \ldots, \mathbf{v}_p) = (-1)^{p+1} t_1 p! A^T \Delta(\boldsymbol{\xi})^{-1}\mathbf{w}$$

where $\mathbf{w}$ is a vector whose ith coordinate is

$$w_i = \left(\frac{\mathbf{a}_i^T \mathbf{v}_1}{\mathbf{a}_i^T \boldsymbol{\xi} - b_i}\right) \cdots \left(\frac{\mathbf{a}_i^T \mathbf{v}_p}{\mathbf{a}_i^T \boldsymbol{\xi} - b_i}\right).$$

The assumption for computing the norm of function $f'(\boldsymbol{\xi})^{-1}D^p f(\boldsymbol{\xi})$ is that $\nu(\mathbf{v}_j) \leq 1$, i.e., $\|\Delta(\boldsymbol{\xi})^{-1}A\mathbf{v}_j\|_2 \leq 1$. Examining the ith coordinate of the vector $\Delta(\boldsymbol{\xi})^{-1}A\mathbf{v}_j$ we conclude that $|\mathbf{a}_i^T \mathbf{v}_j| \leq \mathbf{a}_i^T \boldsymbol{\xi} - b_i$. We apply this inequality to all factors in the previous formula for w_i except the first factor, yielding

$$|w_i| \leq \left|\frac{\mathbf{a}_i^T \mathbf{v}_1}{\mathbf{a}_i \boldsymbol{\xi} - b_i}\right|.$$

We want a bound on $\|\mathbf{w}\|_2$. The sum of squares of the right-hand sides of the preceding inequality would be a bound on $\mathbf{w}^T\mathbf{w}$. But this sum of squares is no more than 1 since $\nu(\mathbf{v}_1) \le 1$. Therefore $\|\mathbf{w}\|_2 \le 1$. Then we compute:

$$\begin{aligned}&\nu((1/p!)f'(\boldsymbol{\xi})^{-1}t_1 p! A^T\Delta(\boldsymbol{\xi})^{-1}\mathbf{w}) = \\ &\qquad \|\Delta(\boldsymbol{\xi})^{-1}A(H + t_1A^T\Delta(\boldsymbol{\xi})^{-2}A)^{-1}t_1A^T\Delta(\boldsymbol{\xi})^{-1}\mathbf{w}\|_2.\end{aligned}$$

Now, we use the same argument as in the proof of Lemma 3.6 to conclude that this quantity is bounded by 1. In particular we observe that the norm can be split into two parts, namely the 2-norm of the long matrix expression multiplied by the 2-norm of $\mathbf{w}$. We use the same technique as earlier for the matrix, and we already know that $\|\mathbf{w}\|_2 \le 1$. ∎

We apply this lemma to the present proof as follows. We take $\boldsymbol{\xi}$ in the lemma to be $\boldsymbol{\xi}(t)$ and f to be $\chi(\cdot, t^+)$. By considering all integer values of $p \ge 2$, we get a bound of 1 on γ in Theorem 3.3.

We have upper bounds for both β and γ. We need to choose δ so that $\beta \le \delta/2 \le 1/(40\gamma)$. Given our bound on γ, this is only possible if we can make $\beta \le 1/40$. Since we already showed that $\beta \le \sqrt{m}\cdot|t - t^+|/t^+$ and we know that $t^+ = \alpha t$, then we should choose α to satisfy

$$\frac{t - \alpha t}{\alpha t}\sqrt{m} \le \frac{1}{40}.$$

Solving this gives

$$\alpha \ge 1 - \frac{1}{1 + 40\sqrt{m}}. \tag{3.16}$$

We can assume for now that α is chosen exactly equal to the right-hand side. Given our bounds for β and γ, we must choose $\delta = 1/20$.

Observe that the root of $\chi(\cdot, t^+)$ is $\boldsymbol{\xi}(t^+)$. We take the center of the ball in the theorem to be $\boldsymbol{\xi}(t)$, and we take the norm in the theorem to be $\nu(\cdot, t)$.

The theorem has another precondition, namely, the ball of radius 4δ must be contained in U. The statement to prove is that a point $\mathbf{y}$ satisfying $\nu(\mathbf{y} - \boldsymbol{\xi}(t), t) \le 1/5$ is interior to Π. This is proved by the same argument used at equation (3.13).

The preconditions of the theorem are now satisfied, so we look at its conclusions. The theorem says that $\boldsymbol{\xi}(t^+)$ is within $3\beta/2$, i.e., within $3/80$, of $\boldsymbol{\xi}(t)$ in the $\nu(\cdot, t)$ norm. Next, the theorem says that the result of taking the Newton step from $\mathbf{x}$, namely, the point $\mathbf{x}^+$, is distance no more than $3.5(1/4)\delta$ from $\boldsymbol{\xi}(t^+)$. This comes out to $3.5/80$. We want to translate this to the $\nu(\cdot, t^+)$ norm. We characterize norm translation with a lemma.

LEMMA 3.8. *Suppose* $\boldsymbol{\xi}_1, \boldsymbol{\xi}_2$ *are vectors in* $\operatorname{Int}\Pi$. *For* $j = 1, 2$, *define norms*

$$\nu_j(\mathbf{x}) = \|\Delta(\boldsymbol{\xi}_j)^{-1}A\mathbf{x}\|_2.$$

Suppose that

$$\nu_1(\boldsymbol{\xi}_1 - \boldsymbol{\xi}_2) \leq s$$

with $s < 1$. *Then*

$$\nu_2(\mathbf{v}) \leq \frac{1}{1-s} \cdot \nu_1(\mathbf{v})$$

for an arbitrary $\mathbf{v} \in \mathbb{R}^n$.

PROOF. We do a computation:

$$\begin{aligned}\nu_2(\mathbf{v}) &= \|\Delta(\boldsymbol{\xi}_2)^{-1} A\mathbf{v}\|_2 \\ &= \|\Delta(\boldsymbol{\xi}_2)^{-1}\Delta(\boldsymbol{\xi}_1)\Delta(\boldsymbol{\xi}_1)^{-1} A\mathbf{v}\|_2 \\ &\leq \|\Delta(\boldsymbol{\xi}_2)^{-1}\Delta(\boldsymbol{\xi}_1)\|_2 \cdot \nu_1(\mathbf{v}).\end{aligned}$$

Therefore, we need a bound for the norm of the preceding diagonal matrix product. The 2-norm of a diagonal matrix is its largest absolute entry, so we have to put a bound on a typical element of this matrix. The ith diagonal entry of the product is

$$\frac{\mathbf{a}_i^T \boldsymbol{\xi}_1 - b_i}{\mathbf{a}_i^T \boldsymbol{\xi}_2 - b_i}.$$

This fraction is always positive. Let r be its reciprocal, i.e.,

$$\begin{aligned}r &= \frac{\mathbf{a}_i^T \boldsymbol{\xi}_2 - b_i}{\mathbf{a}_i^T \boldsymbol{\xi}_1 - b_i} \\ &= 1 - \frac{\mathbf{a}_i^T(\boldsymbol{\xi}_1 - \boldsymbol{\xi}_2)}{\mathbf{a}_i^T \boldsymbol{\xi}_1 - b_i}.\end{aligned}$$

The fraction in this last expression is bounded above by $\nu_1(\boldsymbol{\xi}_1 - \boldsymbol{\xi}_2)$ by definition, and we have assumed that this is no more than s. Therefore, $r \geq 1 - s$. Therefore, the entries of the matrix $\Delta(\boldsymbol{\xi}_2)^{-1}\Delta(\boldsymbol{\xi}_1)$ are at most $1/(1-s)$. This proves the lemma. ∎

For the problem at hand, we apply the lemma with $\mathbf{v} = \mathbf{x}^+ - \boldsymbol{\xi}(t^+)$. We know that the $\nu(\cdot, t)$ norm of this vector is at most 3.5/80. Moreover, we know that the $\nu(\cdot, t)$ norm of $\boldsymbol{\xi}(t^+) - \boldsymbol{\xi}(t)$ is at most 3/80. Since $1/(1-3/80)$ is 80/77,

$$\nu(\mathbf{x}^+ - \boldsymbol{\xi}(t^+), t^+) \leq (80/77)(3.5/80),$$

and the right-hand side simplifies to $1/22$.

This completes the proof that $\mathbf{x}^+$ is a good approximation to $\boldsymbol{\xi}(t^+)$. ∎

3.2.2. *Finding a global minimum*

Theorem 3.2 tells us that the sequence $\mathbf{x}^{(k)}$ stays close in some manner to $\boldsymbol{\xi}(t^{(k)})$ as $t^{(k)}$ gets small. We already know that $\boldsymbol{\xi}(t)$ has an accumulation point at $\mathbf{x}^*$, the global minimum. We now argue that after a certain number of iterations, we can step to an exact global minimum. In order to make this claim, we use a rounding argument; this type of argument was used by Hačijan [1979] to analyze the ellipsoid algorithm. In the following lemmas we are careful about counting rows with only one nonzero entry. If these rows were treated the same as other rows, then the resulting estimates for the algorithm would have a dependence on $m \lg m$ or m^2 in the factor L, which is not appropriate.

LEMMA 3.9. *Let W be an $n \times n$ nonsingular integer matrix. Suppose that no more than f rows of W have two or more nonzero entries. Suppose that ℓ_0 bits are required to write the entries of W. Then*

$$\|W^{-1}\|_\infty < 2^\ell$$

where $\ell = \ell_0 + f^2$.

PROOF. From Lemma 3.1 and the remarks following, we know that if $\mathbf{x}$ is the solution to $W\mathbf{x} = \mathbf{b}$, where $\mathbf{b}$ is a vector whose entries are ± 1, then every entry of $\mathbf{x}$ can be expressed as a ratio of two determinants each of which is less than 2^ℓ. Therefore, the entries of $\mathbf{x}$ are of magnitude less than 2^ℓ. The ∞-norm of the matrix W^{-1} is defined to be the maximum ∞-norm of $W^{-1}\mathbf{b}$ chosen over all vectors $\mathbf{b}$ whose entries are ± 1. ∎

The following lemma will be used four times in the remainder of the chapter, so it is important to understand its statement. The lemma may be summarized as follows. If there is a solution to $W\mathbf{x} = \mathbf{c} + \epsilon$ and $A\mathbf{x} \geq \mathbf{b}$ where $W, \mathbf{c}, A, \mathbf{b}$ have integer entries, then there is a solution to $W\mathbf{x} = \mathbf{c}$, $A\mathbf{x} \geq \mathbf{b}$. This is true provided that ϵ is small enough. The exact solution can be constructed from the approximate solution.

LEMMA 3.10. *Suppose $W \in \mathbb{R}^{p\times p}$, $A \in \mathbb{R}^{m\times p}$, $\mathbf{c} \in \mathbb{R}^p$ and $\mathbf{b} \in \mathbb{R}^m$ all have integer entries. Suppose the system $W\mathbf{x} - \mathbf{c} = \epsilon$, $A\mathbf{x} \geq \mathbf{b}$ has a solution $\mathbf{x}_0$, where ϵ lies in $\mathbb{R}^p$. Suppose that any $p \times p$ submatrix of*

$$\begin{pmatrix} W & \mathbf{c} \\ A & \mathbf{b} \end{pmatrix}$$

requires no more than ℓ_0 bits to write. Let g be the following quantity: g is an upper bound on the rank of

$$\begin{pmatrix} W \\ A \end{pmatrix}$$

after rows with fewer than two nonzero entries are deleted. Suppose $\|\epsilon\|_\infty \leq 2^{-3\ell}$, where $\ell = \ell_0 + g^2$.

Then there exists a solution $\mathbf{x}_1$ to the system $W\mathbf{x} - \mathbf{c} = \mathbf{0}$, $A\mathbf{x} \geq \mathbf{b}$.

PROOF. We will prove this by induction on the dimension of the null space of W. Let this dimension be s.

Base case, $s = 0$. Then W is nonsingular. This means that at most g rows of W have more than one nonzero entry (because if $g + 1$ rows of W had more than one nonzero entry, then the rank of these rows must be at least $g + 1$ since W is full rank, contradicting the definition of g).

Let $\mathbf{d} = W^{-1}\epsilon$. Then we have

$$\|\mathbf{d}\|_\infty \le \|W^{-1}\|_\infty \cdot \|\epsilon\|_\infty$$

and the right-hand side is less than $2^\ell \cdot 2^{-3\ell}$. Since $W\mathbf{x}_0 - \mathbf{c} = W\mathbf{d}$, we can take $\mathbf{x}_1 = \mathbf{x}_0 - \mathbf{d}$ to solve $W\mathbf{x}_1 = \mathbf{c}$.

We know that

$$\|\mathbf{x}_0 - \mathbf{x}_1\|_\infty < 2^{-2\ell}.$$

But we also know that the entries of $\mathbf{x}_1$ have a common denominator less than 2^ℓ. This is because $\mathbf{x}_1$ is the solution to the nonsingular integer system $W\mathbf{x}_1 = \mathbf{c}$. This means that the rational number $b_i - \mathbf{a}_i^T\mathbf{x}_1$ also has denominator less than 2^ℓ since $(\mathbf{a}_i^T, b_i)$ are integers.

Suppose (for a contradiction) that $\mathbf{a}_i^T\mathbf{x}_1 < b_i$. Then the difference $b_i - \mathbf{a}_i^T\mathbf{x}_1$ is more than $2^{-\ell}$. This means that

$$\begin{aligned} b_i - \mathbf{a}_i^T\mathbf{x}_0 &\ge (b_i - \mathbf{a}_i^T\mathbf{x}_1) + \mathbf{a}_i^T(\mathbf{x}_1 - \mathbf{x}_0) \\ &> 2^{-\ell} - p\left(\|\mathbf{a}_i\|_\infty \cdot \|\mathbf{x}_0 - \mathbf{x}_1\|_\infty\right) \\ &> 2^{-\ell} - p \cdot 2^{\ell_0} 2^{-2\ell} \\ &> 0. \end{aligned}$$

This chain of inequalities has led to a contradiction since we assumed that $A\mathbf{x}_0 \ge \mathbf{b}$.

Induction step, $s > 0$. Since W is singular, without loss of generality row 1 of W can be expressed as a linear combination of the other rows. This means that there is a vector $\mathbf{v}$ whose first coordinate is nonzero such that $\mathbf{v}^T W = \mathbf{0}$. By Cramer's rule, we can assume that $\mathbf{v}$ is an integer vector whose largest entry has magnitude at most $2^{2\ell}$.

Since W is singular, there is a vector $\mathbf{d}$ not equal to zero such that $W\mathbf{d} = \mathbf{0}$. Consider the point $\mathbf{x}_0' = \mathbf{x}_0 + t\mathbf{d}$. We can increase or decrease t until either some inequality $\mathbf{a}_i^T\mathbf{x} \ge b_i$ becomes satisfied as an equation or until a coordinate of $\mathbf{x}_0'$ becomes 0. In either case, we still have that $W\mathbf{x}_0' - \mathbf{c} = \epsilon$ and $A\mathbf{x}_0' \ge \mathbf{b}$. Moreover, we can assume that $\mathbf{a}_i^T\mathbf{d} \ne 0$ in the first case, because otherwise the constraint does not change from inactive to active as we move along $\mathbf{d}$. In the second case we can assume that $\mathbf{e}_i^T\mathbf{d} \ne 0$, where $\mathbf{e}_i$ is the ith column of the identity matrix. It may be the case that $t = 0$.

In the first case, we replace the first row of $(W, \mathbf{c})$ with $(\mathbf{a}_i^T, b_i)$ and delete the inequality $\mathbf{a}_i^T\mathbf{x} \ge b_i$ from $A\mathbf{x} \ge \mathbf{b}$. In the second case we replace the first row of $(W, \mathbf{c})$ with $(\mathbf{e}_i^T, 0)$. In both cases we replace the first entry of $\boldsymbol{\epsilon}$ with 0 to yield $\boldsymbol{\epsilon}'$. Call this modified system $(W', \mathbf{c}', \boldsymbol{\epsilon}', A', \mathbf{b}')$. The matrix W' has null space dimension at most $s-1$, because the first row of W' cannot be expressed in terms of the other rows (the other rows are orthogonal to $\mathbf{d}$; the first row is not).

We can apply the lemma by induction to $(W', \mathbf{c}', \boldsymbol{\epsilon}', A', \mathbf{b}')$ with $\mathbf{x}_0'$ as the starting point (notice that $W'\mathbf{x}_0' - \mathbf{c}' = \boldsymbol{\epsilon}'$ and $A'\mathbf{x}_0' \geq \mathbf{b}'$). We need to compute the quantity g' of the modified system. If we are in the case that a row of W was replaced by a row of A, then it is clear that we can take $g' = g$ for the lemma. Similarly in the case that a row of W was replaced by $\mathbf{e}_i^T$, the variable g is not increased because the new row has one nonzero entry. A similar analysis applies to ℓ_0.

We claim that any solution to $W'\mathbf{x} - \mathbf{c}' = \mathbf{0}$ will also be a solution to $W\mathbf{x} - \mathbf{c} = \mathbf{0}$, which would prove the lemma. In particular, we need to verify that the first equation of $W\mathbf{x}_1 = \mathbf{c}$ will hold since this is the equation dropped when we applied the lemma inductively.

Notice that

$$\mathbf{v}^T W \mathbf{x}_0 - \mathbf{v}^T \mathbf{c} = \mathbf{v}^T \boldsymbol{\epsilon}.$$

The left hand side is equal to $-\mathbf{v}^T\mathbf{c}$, an integer, and the right-hand side has absolute value at most $p2^{-\ell}$ since $\|\boldsymbol{\epsilon}\|_2 \leq \sqrt{p}2^{-3\ell}$ and $\|\mathbf{v}\|_2 \leq \sqrt{p}2^{2\ell}$. Thus, the left hand side is an integer and the right-hand side has magnitude less than 1. The only way this is possible is for both sides to be zero. Thus, $\mathbf{v}^T\mathbf{c} = 0$, so the first equation of the system given by $(W, \mathbf{c})$ is in fact implied by the other equations. Therefore, since the other equations are solved, the lemma follows. ∎

Now, we return to the Renegar-Shub algorithm. We halt at stage k when

$$t^{(k)} \leq \frac{19 \cdot 2^{-12L}}{84m \cdot \|A^T\|_\infty}.$$

(Recall that L was defined by (3.8).) The motivation for this fraction on the right-hand side will become apparent from the calculations in this section. Denote $\boldsymbol{\xi}, \mathbf{x}, t$ at this iteration by $\boldsymbol{\xi}_a, \mathbf{x}_a, t_a$.

Notice that we have the identity:

$$H\boldsymbol{\xi}_a + \mathbf{c} - t_a A^T \Delta(\boldsymbol{\xi}_a)^{-1}\mathbf{e} = \mathbf{0}. \tag{3.17}$$

Let U be the set of indices i such that

$$\mathbf{a}_i^T \mathbf{x}_a - b_i \leq \frac{(19/20)2^{-6L}}{2\sqrt{m}}.$$

Let $(A_U, \mathbf{b}_U)$ denote the submatrix of $(A, \mathbf{b})$ whose rows are chosen from U. Since $\mathbf{x}_a$ is a good approximation to $\boldsymbol{\xi}_a$ we know by (3.13) that for $i \in U$,

$$\mathbf{a}_i^T \boldsymbol{\xi}_a - b_i \leq 2^{-6L}/(2\sqrt{m}), \tag{3.18}$$

and for $i \notin U$,

$$\mathbf{a}_i^T \boldsymbol{\xi}_a - b_i \geq (19/21)2^{-6L}/(2\sqrt{m}). \tag{3.19}$$

Let $\boldsymbol{\mu}_a$ be the entries of vector $t_a\Delta(\boldsymbol{\xi}_a)^{-1}\mathbf{e}$ indexed by U. Then we can write

$$\begin{aligned} H\boldsymbol{\xi}_a + \mathbf{c} - A_U^T\boldsymbol{\mu}_a &= \boldsymbol{\epsilon}', \\ A_U\boldsymbol{\xi}_a - \mathbf{b}_U &= \boldsymbol{\epsilon}'', \\ A\boldsymbol{\xi}_a &\geq \mathbf{b}, \\ \boldsymbol{\mu}_a &\geq \mathbf{0} \end{aligned} \tag{3.20}$$

where $\boldsymbol{\epsilon}'$ and $\boldsymbol{\epsilon}''$ are vectors that we now analyze.

First, note that $\|\boldsymbol{\epsilon}''\|_\infty \leq 2^{-6L}$ by (3.18). (In fact, it is no more than $2^{-6L}/(2\sqrt{m})$; we need the extra factor later on). For the other equation, we observe from (3.17) and by choice of $\boldsymbol{\mu}_a$ that

$$\boldsymbol{\epsilon}' = t_a \sum_{i \notin U} \frac{\mathbf{a}_i}{\mathbf{a}_i^T\boldsymbol{\xi}_a - b_i}.$$

Note that $\|\boldsymbol{\epsilon}'\|_\infty$ is at most

$$\frac{42t_a\sqrt{m}}{19 \cdot 2^{-6L}} \cdot \|A^T\|_\infty;$$

this follows from (3.19). Thus, $\|\boldsymbol{\epsilon}'\|_\infty \leq 2^{-6L}/(2\sqrt{m})$ by choice of t_a, and in particular, this norm is no more than 2^{-6L}.

Accordingly, we apply Lemma 3.10 to (3.20) to conclude that there exists a pair of vectors $(\mathbf{x}^*, \boldsymbol{\mu}^*)$ such that the following conditions hold.

$$\begin{aligned} H\mathbf{x}^* + \mathbf{c} - A_U^T\boldsymbol{\mu}^* &= \mathbf{0}, \\ A_U\mathbf{x}^* - \mathbf{b}_U &= \mathbf{0}, \\ A\mathbf{x}^* &\geq \mathbf{b}, \\ \boldsymbol{\mu}^* &\geq \mathbf{0}. \end{aligned} \tag{3.21}$$

Note that for the lemma $p = n + |U|$. The number ℓ_0 in the lemma is at most $2L_0$. Finally, the quantity g is at most $2n$. This is because the rank of the constraints of (3.21), not counting the constraints $\boldsymbol{\mu}^* \geq \mathbf{0}$, is at most $2n$. Thus, ℓ in the lemma is at most $2L$. System (3.21) is exactly the system of KKT conditions, so the point $\mathbf{x}^*$ is a global minimum for the original quadratic program (3.7). Let

$$W = \begin{pmatrix} H & -A_U^T \\ A_U & 0 \end{pmatrix}.$$

Note that W can be determined from $\mathbf{x}_a$; in particular, W depends only on U. If W is nonsingular, then we can construct the solution to (3.21) simply by solving the relevant system of linear equations. This is the nondegenerate case—in this case the only information from the iterative method needed to compute an optimum is the index set U.

Next, we address the degenerate case. In this case the matrix W is singular. Lemma 3.10 is constructive when W is singular, but only if we have the vector $\mathbf{x}_0$ of the lemma (i.e., the pair $(\boldsymbol{\xi}_a, \boldsymbol{\mu}_a)$) on hand, which is not the case. Accordingly, some other approach is needed.

In this regard, define $\boldsymbol{\mu}'_a$ to be the vector of entries of $t_a\Delta(\mathbf{x}_a)^{-1}\mathbf{e}$ indexed by U. Consider the following unconstrained minimization problem: Find $\mathbf{x}_b$ and $\boldsymbol{\mu}_b$ to minimize

$$(19T_1)^2 + (20T_2)^2 + (2 \cdot 2^{6L}T_3)^2 + (2 \cdot 2^{6L}T_4)^2 \tag{3.22}$$

where

$$\begin{aligned}
T_1 &= \|\Delta(\mathbf{x}_a)^{-1}A(\mathbf{x}_b - \mathbf{x}_a)\|_2, \\
T_2 &= \left(\sum_{i\in U} \frac{(\mu'_{a,i} - \mu_{b,i})^2}{(\mu'_{a,i})^2}\right)^{1/2}, \\
T_3 &= \|H\mathbf{x}_b + \mathbf{c} - A_U^T\boldsymbol{\mu}_b\|_2, \\
T_4 &= \|A_U\mathbf{x}_b - \mathbf{b}_U\|_2.
\end{aligned}$$

This is a linear least squares problem that is equivalent to a nonsingular system of linear equations. It is nonsingular because T_1^2 is a full-rank quadratic form of $\mathbf{x}_b$ and T_2^2 is a full-rank quadratic form of $\boldsymbol{\mu}_b$. Note that all the coefficients of this linear least squares system are derived from $\mathbf{x}_a$ and t_a, two known quantities.

We claim that the objective function of this unconstrained minimization problem evaluated at $(\boldsymbol{\xi}_a, \boldsymbol{\mu}_a)$ (a point that our algorithm has not computed) is bounded by 4. In particular, it is left as an exercise (Exercise 3.5) that each of the four terms of (3.22) is at most 1 if we substitute $(\boldsymbol{\xi}_a, \boldsymbol{\mu}_a)$ for $(\mathbf{x}_b, \boldsymbol{\mu}_b)$ above. For the first and second term, this follows because $\mathbf{x}_a$ is a good approximation to $\boldsymbol{\xi}_a$. For the third and fourth terms, this follows from the estimates derived earlier for ϵ' and ϵ'' (multiplying by $\sqrt{n}$ or $\sqrt{m}$ to convert to the 2-norm).

Therefore, the minimum of the above system will be achieved at a point $(\mathbf{x}_b, \boldsymbol{\mu}_b)$ in which no term of the objective function before squaring is more than 2. Let us analyze this pair of vectors. First,

$$T_1 \leq 2/19$$

so that $\mathbf{a}_i^T\mathbf{x}_b - b_i$ cannot be more than $2/19$ times smaller or larger than $\mathbf{a}_i^T\mathbf{x}_a - b_i$. In particular, this means that $A\mathbf{x}_b \geq \mathbf{b}$. Second, the bound $T_2 \leq 1/10$ tells us that

$$|\mu_{b,i} - \mu'_{a,i}| \leq \mu'_{a,i}/10.$$

Since $\boldsymbol{\mu}'_a$ is a positive vector, this shows that $\boldsymbol{\mu}_b$ is also positive. The bounds on T_3 and T_4 tell us that

$$\|H\mathbf{x}_b + \mathbf{c} - A_U^T\boldsymbol{\mu}_b\|_2 \le 2^{-6L}$$

and

$$\|A_U\mathbf{x}_b - \mathbf{b}_U\|_2 \le 2^{-6L}.$$

These same bounds must hold for the ∞-norm.

This yields a system very much like (3.20), except in terms of $\mathbf{x}_b$ and $\boldsymbol{\mu}_b$. Therefore, we can apply Lemma 3.10 to find a solution to (3.21), a global minimum. In this case $\mathbf{x}_b$ and $\boldsymbol{\mu}_b$ have been calculated so the lemma may be applied in a constructive fashion.

We have now demonstrated that we can find a global minimum to the original quadratic program using the Renegar-Shub algorithm. Three topics still need to be discussed, namely, initializing the algorithm, bounding the number of arithmetic operations, and bounding the number of digits. These topics are the subject of the upcoming three subsections.

3.2.3. *Initializing the algorithm*

Next, we address the problem of initializing the algorithm. Recall that we have made four assumptions about how the algorithm starts, namely, the feasible region is nonempty and compact, and that A has rank n. In addition, we have assumed that we have a strictly interior feasible point $\mathbf{x}^{(0)}$. This initial feasible point must be within 1/20 of $\boldsymbol{\xi}(t^{(0)})$ in the $\nu(\cdot, t^{(0)})$ norm. In the following theorem, we show how to modify the original problem so that all these assumptions can be satisfied.

THEOREM 3.4. *Suppose* (3.7) *is feasible and has a global minimum. Then the system*

$$\begin{aligned} \textit{minimize} \quad & \tfrac{1}{2}\mathbf{x}^T H\mathbf{x} + \mathbf{c}^T\mathbf{x} + \theta y \\ \textit{subject to} \quad & A\mathbf{x} + \mathbf{f}y \ge \mathbf{b}, \\ & l_i \le x_i \le u_i \\ & 0 \le y \le r \end{aligned} \tag{3.23}$$

has a global minimum $(\mathbf{x}^*, y^*)$ *such that* $\mathbf{x}^*$ *is a global minimum of* (3.7). *Moreover, this problem has a compact feasible region, the constraint matrix has rank* $n+1$, *and the point* $(\mathbf{0}, y_0)$ *is within* $1/20$ *of* $\boldsymbol{\xi}(t^{(0)})$ *in the norm* $\nu(\cdot, t^{(0)})$. *Here,* $\mathbf{f}$ *is a vector of positive integers,* θ, $t^{(0)}$ *and* y_0 *are large integers, and* r, $-l_i$,u_i *are large rational numbers (see below for the values of these quantities).*

PROOF. Note immediately that the feasible region for (3.23) is compact, and the constraint matrix has rank $n+1$. Both of these claims follow from the presence of the constraints that are bounds on the variables.

Below we will choose l_i less than -2^{2L} and u_i greater than 2^{2L}. If the original problem (3.7) had a global optimum, then it also has one $\mathbf{x}_0$ such that $\|\mathbf{x}_0\|_\infty \le 2^{2L}$. (The problem of detecting whether the original problem has a minimum is Exercise 3.8.) The point $(\mathbf{x}_0, 0)$ is feasible for (3.23) with the same value of the objective function. Indeed, any point of the form $(\mathbf{x}, 0)$ feasible for (3.23) will have $\mathbf{x}$ feasible for (3.7) with the same value of the objective function.

Now, we want to argue that if θ is chosen large enough, then any optimal solution $(\mathbf{x}^*, y^*)$ to (3.23) have $y^* = 0$, unless the original problem was infeasible. If $y^* = 0$, then a global minimum for the original problem (3.7) can be extracted from $(\mathbf{x}^*, y^*)$. We state this as a lemma.

LEMMA 3.11. *Let $(\mathbf{x}^*, y^*)$ be an optimal solution to (3.23). If y^* is positive, then the original system is infeasible, provided θ is chosen large enough.*

PROOF. Let $(\mathbf{x}^*, y^*)$ be a global minimum with $y^* > 0$. We assume that $y^* < r$; the other case is similar and is left to the reader. We write out the KKT conditions for $(\mathbf{x}^*, y^*)$ under these assumptions: There exist index sets U, V, W and vectors $\boldsymbol{\mu} \in \mathbb{R}^{|U|}$, $\boldsymbol{\sigma}, \boldsymbol{\tau} \in \mathbb{R}^n$ such that

$$
\begin{aligned}
H\mathbf{x}^* + \mathbf{c} &= A_U^T\boldsymbol{\mu} + \boldsymbol{\sigma} - \boldsymbol{\tau}, \\
\mathbf{f}_U^T\boldsymbol{\mu} &= \theta, \\
A_U\mathbf{x}^* + \mathbf{f}_U y^* - \mathbf{b}_U &= \mathbf{0}, \\
A\mathbf{x}^* + \mathbf{f}y^* - \mathbf{b} &\ge \mathbf{0}, \\
x_i^* &= l_i \quad \text{for } i \in W, \\
\sigma_i &= 0 \quad \text{for } i \notin W, \\
x_i^* &= u_i \quad \text{for } i \in V, \\
\tau_i &= 0 \quad \text{for } i \notin V, \\
\boldsymbol{\mu} &\ge \mathbf{0}, \\
y^* &\in (0, r), \\
\mathbf{x}_i^* &\in [l_i, u_i] \quad \text{for } i = 1, \ldots, n, \\
\boldsymbol{\sigma} &\ge \mathbf{0}, \\
\boldsymbol{\tau} &\ge \mathbf{0}.
\end{aligned}
$$

We use the notation $(A_U, \mathbf{b}_U)$ to denote the rows of $(A, \mathbf{b})$ indexed by U. Set $\boldsymbol{\mu}' = \boldsymbol{\mu}/\theta$, $\boldsymbol{\sigma}' = \boldsymbol{\sigma}/\theta$, $\boldsymbol{\tau}' = \boldsymbol{\tau}/\theta$ and substitute. Then we can rewrite the above system

$$
\begin{aligned}
A_U^T\boldsymbol{\mu}' + \boldsymbol{\sigma}' - \boldsymbol{\tau}' &= \boldsymbol{\epsilon}, \\
\mathbf{f}_U^T\boldsymbol{\mu}' &= 1, \\
A_U\mathbf{x}^* + \mathbf{f}_U y^* - \mathbf{b}_U &= \mathbf{0}, \\
A\mathbf{x}^* + \mathbf{f}y^* - \mathbf{b} &\ge \mathbf{0}, \\
x_i^* &= l_i \quad \text{for } i \in W,
\end{aligned}
$$

$$\begin{aligned}
\sigma_i' &= 0 \quad \text{for } i \notin W, \\
x_i^* &= u_i \quad \text{for } i \in V, \\
\tau_i' &= 0 \quad \text{for } i \notin V, \\
\boldsymbol{\mu} &\geq \mathbf{0} \\
y^* &\in (0, r), \\
\mathbf{x}_i^* &\in [l_i, u_i] \quad \text{for } i = 1, \ldots, n \\
\boldsymbol{\sigma}' &\geq \mathbf{0}, \\
\boldsymbol{\tau}' &\geq \mathbf{0},
\end{aligned} \tag{3.24}$$

where $\boldsymbol{\epsilon} = (H\mathbf{x}^* + \mathbf{c})/\theta$. Suppose we choose θ large enough so that

$$\|\boldsymbol{\epsilon}\|_\infty \leq 2^{-3\ell} \tag{3.25}$$

where ℓ is the size of system (3.24) (to be determined below) not counting ϵ. Note that variable g for system (3.24) in Lemma 3.10 is at most $2n+2$; this can be checked with a careful counting. Inequality (3.25) will always hold provided that θ is at least $|2^{3\ell+3L} l_i|$ and at least $2^{3\ell+3L} u_i$ for all i. Therefore we can apply Lemma 3.10 to discover a solution $(\mathbf{x}'', y'', \boldsymbol{\mu}'', \boldsymbol{\sigma}'', \boldsymbol{\tau}'')$ to the system

$$\begin{aligned}
A_U^T \boldsymbol{\mu}'' + \boldsymbol{\sigma}'' - \boldsymbol{\tau}'' &= \mathbf{0}, \\
\mathbf{f}_U^T \boldsymbol{\mu}'' &= 1, \\
A_U \mathbf{x}'' + \mathbf{f}_U y'' - \mathbf{b}_U &= \mathbf{0}, \\
A\mathbf{x}'' + \mathbf{f} y'' - \mathbf{b} &\geq \mathbf{0}, \\
x_i'' &= l_i \quad \text{for } i \in W, \\
\sigma_i'' &= 0 \quad \text{for } i \notin W, \\
x_i'' &= u_i \quad \text{for } i \in V, \\
\tau_i'' &= 0 \quad \text{for } i \notin V, \\
\boldsymbol{\mu}'' &\geq \mathbf{0}, \\
y'' &\in [0, r], \\
\mathbf{x}_i'' &\in [l_i, u_i] \quad \text{for } i = 1, \ldots, n, \\
\boldsymbol{\sigma}'' &\geq \mathbf{0}, \\
\boldsymbol{\tau}'' &\geq \mathbf{0}.
\end{aligned}$$

We claim that the existence of such a $(\boldsymbol{\mu}'', \boldsymbol{\tau}'', \boldsymbol{\sigma}'')$ contradicts the feasibility of the system $A\mathbf{x} \geq \mathbf{b}$, $\|\mathbf{x}\|_\infty \leq 2^{2L}$, which in turn contradicts the feasibility of the original problem (3.7). In particular, suppose there were a vector $\mathbf{x}$ satisfying $A\mathbf{x} \geq \mathbf{b}$, $\|\mathbf{x}\|_\infty \leq 2^{2L}$. This means $A_U\mathbf{x} \geq \mathbf{b}_U$. Take the inner product of both sides with $\boldsymbol{\mu}''$ (a nonnegative vector). Substituting $(\boldsymbol{\mu}'')^T A_U = (\boldsymbol{\tau}'' - \boldsymbol{\sigma}'')^T$ from the above system, we get

$$(\boldsymbol{\tau}'' - \boldsymbol{\sigma}'')^T \mathbf{x} \geq (\boldsymbol{\mu}'')^T \mathbf{b}_U.$$

Next, we observe that $\mathbf{b}_U = A_U\mathbf{x}^* + \mathbf{f}_U y^*$. Substituting and using the above formulas for $(\boldsymbol{\mu}'')^T\mathbf{f}_U$ and $(\boldsymbol{\mu}'')^T A_U$ gives:

$$(\boldsymbol{\tau}'' - \boldsymbol{\sigma}'')^T\mathbf{x} \geq (\boldsymbol{\tau}'' - \boldsymbol{\sigma}'')^T\mathbf{x}^* + y^*. \tag{3.26}$$

We claim this is impossible. In particular, $y^* > 0$ by assumption. Moreover, one checks that

$$(\tau_i'' - \sigma_i'')x_i \leq (\tau_i'' - \sigma_i'')x_i^*$$

for all i. This is because for any i in which the parenthesized expression is positive, $i \in V$ and $x_i^* = u_i$ (so $x_i^* \geq x_i$). Similarly, for any i in which the parenthesized expression is negative, $x_i^* = l_i \leq x_i$.

Thus, (3.26) is impossible, so the system $A_U\mathbf{x} \geq \mathbf{b}_U$ has no solution, so the original problem was infeasible. ∎

This establishes one part of the Theorem 3.4, namely, that the global optimum of (3.23) will be the same as the global optimum of (3.7). Now we establish the remainder of the theorem, that is, the existence of an initial feasible point.

We take the point $(\mathbf{0}, y_0)$ as our initial feasible point with y_0 chosen to be very large. Clearly this point is feasible and interior since $\mathbf{f}$ is a vector of positive integers. We shall show that this is close to $\boldsymbol{\xi}(t^{(0)})$ if we choose all the parameters properly. We start with $\mathbf{f}$; we choose its ith coordinate as follows:

$$f_i = \begin{cases} b_i & \text{if } b_i > 0, \\ 1 & \text{if } b_i = 0, \text{ and} \\ -b_i & \text{if } b_i < 0. \end{cases}$$

The motivation for this choice is explained below. We take $y_0 = 2^{4L}$. For quadratic program (3.23) the function ∇h evaluated at $(\mathbf{0}, y_0)$ has the following value

$$\begin{aligned} \nabla h((\mathbf{0}, y_0), t^{(0)}) = \begin{pmatrix} \mathbf{c} \\ \theta \end{pmatrix} - t^{(0)} \Bigg[&\sum_{i=1}^{m} \begin{pmatrix} \mathbf{a}_i \\ f_i \end{pmatrix} \cdot \frac{1}{f_i y_0 - b_i} \\ &- \sum_{i=1}^{n} \begin{pmatrix} \mathbf{e}_i \\ 0 \end{pmatrix} \cdot \left(\frac{1}{l_i} + \frac{1}{u_i} \right) \\ &- \begin{pmatrix} \mathbf{0} \\ 1 \end{pmatrix} \cdot \left(\frac{1}{r - y_0} - \frac{1}{y_0} \right) \Bigg]. \end{aligned} \tag{3.27}$$

We want to make the bracketed term of this equation come out to zero. We need to pick all of the parameters in the right order. We address the first

n coordinates of the bracketed sum first, with the goal being to force these entries of the above total to be zero.

The magnitude of each entry of each term of the first summation is bounded by $2^{L_0} \cdot 2^{-4L+1}$ because of bounds on $\|A^T\|_\infty$ and choice of y_0. We next examine the size of the denominators of these rational numbers. Note that by choice of **f**, each denominator $f_i y_0 - b_i$ is either $(y_0 - 1)b_i$, $(y_0 + 1)b_i$, or y_0, depending on the sign of b_i. Therefore, the common denominator of the first summation is at most $y_0(y_0 - 1)(y_0 + 1)b_1 \cdots b_m$ (omitting b_i's that are zero), which is at most 2^{13L} by choice of y_0. This is the purpose of our choice for **f**; if we had chosen **f** to be the vector of all 1's (for example), then this common denominator could be as large as 2^{4mL}. Call this maximum common denominator d; note that it is independent of i.

Next we can select l_i and u_i. We choose them to cancel out the contribution from the first summation in (3.27). Let the ith coordinate in the sum of the first summation be denoted by v_i. We have already argued that v_i is rational of the form j/d where d is at most 2^{13L} and j has magnitude at most $2^{-3L}d$.

Then we choose

$$l_i = \frac{d}{-2^{-3L}d + j/2}$$

and

$$u_i = \frac{d}{2^{-3L}d + j/2}.$$

Notice that

$$\frac{1}{l_i} + \frac{1}{u_i} = j/d$$

so that we get the desired cancellation. Notice also that $-l_i$ and u_i are at least 2^{2L} as required. They are at most 2^{3L+1} in magnitude.

Next we choose r so that the last coordinate entry of the bracketed term above is cancelled. By the same argument as in the previous few paragraphs, we know that the contribution to the last coordinate from the first bracketed sum of (3.27) is a fraction of the form j/d with the same d, and j at most $2^{-3L}d$ in magnitude. Moreover, we know that this fraction is positive since each f_i is positive. Therefore, if we set

$$r = \frac{d}{2^{-4L}d + j} + 2^{4L},$$

then the last coordinate entry is cancelled.

Thus, with these choices,

$$\nabla h((\mathbf{0}, y_0), t^{(0)}) = \begin{pmatrix} \mathbf{c} \\ \theta \end{pmatrix}.$$

Denote this vector as $\hat{\mathbf{c}}$. Next we choose θ. The bound on θ was given by (3.25), where ℓ denotes the size of the system (3.24). If we substitute the values calculated for l_i, u_i, $\mathbf{f}$ and r in (3.24) and clear denominators, we find the largest integer is at most $2^{8L}d$. Thus, ℓ is at most $21L+1$, so we can take $\theta = 2^{69L+4}$.

Finally we have to pick $t^{(0)}$ to make $(\mathbf{0}, y_0)$ a good approximation to $\boldsymbol{\xi}(t^{(0)})$. Define a new norm $\nu_1(\cdot)$ by

$$\nu_1(\mathbf{x}) = \|\Delta(\mathbf{0}, y_0)^{-1}\hat{A}\mathbf{x}\|_2.$$

We use $(\hat{H}, \hat{\mathbf{c}}, \hat{A}, \hat{\mathbf{b}})$ to denote the objective function coefficients and constraint coefficients of the augmented system (3.23). The notation $\Delta(\mathbf{x})$ refers to this augmented system in the present context. Now we want to apply Theorem 3.3 to bound the distance from $(\mathbf{0}, y_0)$ to $\boldsymbol{\xi}(t^{(0)})$ in the norm ν_1. The function used for the theorem is $\nabla h(\cdot, t^{(0)})$, which has its root at $\boldsymbol{\xi}(t^{(0)})$. We apply the theorem with the center of the ball taken to be $(\mathbf{0}, y_0)$. Following the analysis in the proof of Lemma 3.6, we need to bound

$$\beta = \|M(\hat{H} + t^{(0)}M^TM)^{-1}\hat{\mathbf{c}}\|_2 \tag{3.28}$$

where M denotes $\Delta(\mathbf{0}, y_0)^{-1}\hat{A}$. Recall that M^TM is the sum of positive semidefinite matrices of the form

$$\frac{\hat{\mathbf{a}}_i\hat{\mathbf{a}}_i^T}{(\hat{\mathbf{a}}_i^T(\mathbf{0}, y_0) - \hat{b}_i)^2}.$$

Included in M^TM are terms of the form

$$\frac{\mathbf{e}_i\mathbf{e}_i^T}{l_i^2}$$

for $i = 1, \ldots, n$ and a term equal to

$$\frac{\mathbf{e}_{n+1}\mathbf{e}_{n+1}^T}{y_0^2}.$$

This means that M^TM is full rank, and moreover, its smallest eigenvalue is at least 2^{-8L}. Thus, the 2-norm of the parenthesized matrix in (3.28) is at least $2^{-8L}t^{(0)}$.

Moreover, the 2-norm of M is at most 2^{-L} because the largest entry of $\Delta(\mathbf{0}, y_0)^{-1}$ is bounded by 2^{-2L}.

Therefore, since $\|\hat{\mathbf{c}}\|_2$ is at most 2^{69L+5}, we can force β to be less than $1/80$ by choosing $t^{(0)}$ to be 2^{76L+12}.

Next we need to bound γ in Theorem 3.3. We get a bound of 1 as in Lemma 3.7 because ν_1 is defined based on the center of the ball $(\mathbf{0}, y_0)$.

Thus, we apply the theorem with $\delta = 1/40$ to deduce that the distance between $(\mathbf{0}, y_0)$ and $\boldsymbol{\xi}(t^{(0)})$ in the ν_1 norm is at most $3\beta/2$, i.e., at most $3/160$. Then Lemma 3.8 tells us that the distance between these points in the $\nu(\cdot, t^{(0)})$ norm is at most $(160/157)\cdot(3/160)$ which is bounded by $1/20$. ∎

3.2.4. *Running time analysis*

Next, we look into the number of arithmetic operations required by the algorithm. For this subsection and the next subsection, we return to the assumption that the number L is the number of bits to write the entire system being solved, and that the coefficients of the system are $(H, \mathbf{c}, A, \mathbf{b})$.

Finally, we adopt the following notation. In the previous section we stated that we could choose $t^{(0)} = 2^{76L+12}$. From now on, we will assume that $t^{(0)} = 2^{cL}$ for a positive constant c. This will allow us to show how the bounds in this subsection depend on $t^{(0)}$.

Observe that each iteration entails the solution of a system of linear equations, the most expensive step of the solution process. This system, given by

$$\chi'(\mathbf{x}, t) = H + tA^T \Delta(\mathbf{x})^{-2} A,$$

is symmetric positive definite and may be solved with LDL^T factorization. Factorization requires $n^3/3$ arithmetic operations. Forming the matrix product $A^T \Delta(\mathbf{x})^{-2} A$ requires $O(mn^2)$ operations.

Both steps can be reduced to $O(m^{.5}n^2)$ operations on average by using an approximation to $\Delta(\mathbf{x})$ and updating the LDL^T factors with rank-one updates. We omit the discussion of this technique, which is due to Karmarkar [1984] and is explained for the algorithm at hand by Renegar and Shub [1988]. The running time for the special case of linear programming can be brought down even further using fast matrix multiplication techniques, as observed by Vaidya [1989]. Vaidya achieves the best known asymptotic running time bound for linear programming.

Next we consider the question of how many iterations are required. Recall that we need to iterate to find $\mathbf{x}_a$. The vector $\mathbf{x}_a$ is found when $t^{(k)} \leq t^*$, where

$$t^* = \frac{19 \cdot 2^{-12L}}{84m \cdot \|A^T\|_\infty}.$$

Since

$$t^{(k)} = \left(1 - \frac{1}{1 + 40\sqrt{m}}\right)^k t^{(0)}$$

we are left with the problem of solving for k. Taking logs, this problem is

$$k \lg \left(1 - \frac{1}{1 + 40\sqrt{m}}\right) + cL \leq \lg t^*.$$

A lower bound on the right-hand side is $-13L$. Since $\ln x \leq x - 1$ for all positive real numbers x, then $\lg(1 - x) \leq -x \cdot \lg e$ for all numbers $x \in (0, 1)$. So it suffices to solve the problem

$$-k \cdot \frac{1}{1 + 40\sqrt{m}} \leq -(cL + 13L) \lg e.$$

If we bound $1 + 40\sqrt{m}$ with $41\sqrt{m}$ then the inequality is solved by $k = (60c + 769)\sqrt{m}L$.

Thus, at most $(60c + 769)\sqrt{m}L$ iterations are required to find $\mathbf{x}_a$. Note that in practice this bound is a substantial overestimate. The number of iterations for the interior point methods most efficient in practice is usually bounded by 50 on test problems. Moreover, the number of iterations until convergence seems to be nearly independent of m. See Mizuno, Todd, and Ye [1989] for an indication of why the $\sqrt{m}$ factor fails to appear in practice. The upper bound of $O(\sqrt{m}L)$ iterations is the best in the literature.

From $\mathbf{x}_a$ we can step to a global solution, as explained above. Thus, an upper bound on the number of arithmetic operations is $O(m^{1.5}n^2L)$ to find a global minimum. This is a polynomial bound since the length of the input is at least $\max(mn, L)$.

3.2.5. *Polynomial number of digits*

We need to show that the algorithm can be carried out with a polynomial number of digits. So far we have been assuming exact accuracy in all calculations. We address the main iteration of the Renegar-Shub algorithm, since all the other steps involve systems of linear equations which we know how to handle.

The analysis of the number of digits is primarily a theoretical issue, although there is some connection to the numerical stability of practical interior point methods. In particular, there is an indirect relation between the number of digits and the condition number of the Hessian of the potential function; see Exercise 3.6.

Notice that one step of the Renegar-Shub algorithm requires the solution of a system of linear equations. This can be carried out in exact rational arithmetic. The only difficulty is that $\sqrt{m}$, an irrational number, figures into the value of $t^{(k)}$. This is easily solved by replacing

$$\alpha = 1 - \frac{1}{1 + 40\sqrt{m}}$$

with

$$\alpha = 1 - \frac{1}{1 + 40q}$$

where q is the square root of m rounded up to one decimal place. Then α will have a polynomial number of digits, and q is at most $11/10$ of $\sqrt{m}$, so this changes the running time bound by only a small constant factor.

At the end of an iteration, we have $\mathbf{x}^+$ and t^+ which, although rational, could require many more digits than $\mathbf{x}$ and t. If this process continued unchecked, the number of digits for exact arithmetic would grow very rapidly after a few iterations. We solve this problem as follows. At the end of the iteration, we round each coordinate of $\mathbf{x}^+$ to the $(27L + 10)$th bit past the decimal point. Since no entry of $\mathbf{x}$ ever exceeds 2^{2L} in magnitude, this rounding means that at most $O(L)$ digits need to be saved. Let $\mathbf{x}_\sim$ denote the rounded version of this vector.

The rounding for t^+ requires only a few digits. We round t^+ to an approximation $t_\sim$ that satisfies:

$$\left|\frac{t_\sim - t^+}{t_\sim}\right| \leq \frac{1}{1500q}.$$

Note that this rule requires that we round t^+ to a fixed number of significant digits; compare this to $\mathbf{x}^+$, which was rounded to a fixed number of decimal places.

We first need to show that $t_\sim$ is still tending to zero fast enough for the algorithm to converge. In this regard, observe that

$$t^+ \leq t \cdot \left(1 - \frac{1}{41q}\right)$$

and

$$t_\sim \leq t^+ \cdot \left(1 + \frac{1}{1500q - 1}\right)$$

so, multiplying out, we find that

$$t_\sim \leq t \cdot \left(1 - \frac{1}{43q}\right).$$

Thus, the algorithm still converges after $O(\sqrt{m}L)$ iterations.

Let $\boldsymbol{\xi}_\sim$ be $\boldsymbol{\xi}(t_\sim)$. The other main thing we need to show is that

$$\nu(\boldsymbol{\xi}_\sim - \mathbf{x}_\sim, t_\sim) \leq 1/20$$

in order to make Theorem 3.2 work out.

This is proved using the triangle inequality. Let $\boldsymbol{\xi}^+$ denote $\boldsymbol{\xi}(t^+)$. We bound the distance from $\boldsymbol{\xi}_\sim$ to $\boldsymbol{\xi}^+$, the distance from $\boldsymbol{\xi}^+$ to $\mathbf{x}^+$, and finally, the distance from $\mathbf{x}^+$ to $\mathbf{x}_\sim$. First we work in the $\nu(\cdot, t^+)$ norm, and at the end of the argument we convert to the $\nu(\cdot, t_\sim)$ norm. Recall that we proved

$$\nu(\boldsymbol{\xi}^+ - \mathbf{x}^+, t^+) \leq 1/22$$

in the proof of Theorem 3.2. This is one of the three distances that needs to be bounded.

Now we claim that

$$\nu(\mathbf{x}^+ - \mathbf{x}_\sim, t^+) \leq 1/1000,$$

establishing another one of the necessary bounds. To prove this, we first assert the following lemma.

LEMMA 3.12. *For any $t > 0$, the largest diagonal entry of $t\Delta(\boldsymbol{\xi}(t))^{-1}$ is less than* $\max(2^{13L}, 2^{13L}t)$.

PROOF. For the rest of this proof let $\boldsymbol{\xi}$ denote $\boldsymbol{\xi}(t)$. Define

$$\mu_i = \frac{t}{\mathbf{a}_i^T \boldsymbol{\xi} - b_i},$$

i.e., the ith diagonal entry of $t\Delta(\boldsymbol{\xi})^{-1}$. Suppose the lemma is false, i.e., suppose there exists an i such that $\mu_i \geq 2^{13L} t_1$, where $t_1 = \max(t, 1)$. Let U be the set of indices i such that $\mu_i \geq 2^{6L} t_1$. Let $(A_U, \mathbf{b}_U)$ denote the submatrix of $(A, \mathbf{b})$ with rows chosen from U, and let $\boldsymbol{\mu}$ denote the vector of μ_i for $i \in U$. Then we have the system

$$\begin{aligned} H\boldsymbol{\xi} + \mathbf{c} - A_U^T \boldsymbol{\mu} &= \mathbf{v}, \\ A_U \boldsymbol{\xi} - \mathbf{b}_U &= \boldsymbol{\epsilon}, \\ \mathbf{e}^T \boldsymbol{\mu} &\geq 2^{13L} t_1, \\ \boldsymbol{\mu} &\geq \mathbf{0}. \end{aligned}$$

Here, $\mathbf{v}$ is the vector

$$t \sum_{i \notin U} \frac{\mathbf{a}_i}{\mathbf{a}_i^T \boldsymbol{\xi} - b_i},$$

which, by choice of U, has ∞-norm at most $2^{6L} t_1 \|A^T\|_\infty$, i.e., no more than $2^{7L} t_1$. Also, $H\boldsymbol{\xi} + \mathbf{c}$ has norm at most 2^{3L_0}. Finally, $\boldsymbol{\epsilon}$ is a vector with ∞-norm at most $t/(2^{6L} t_1)$, i.e., at most 2^{-6L}.

Substitute $\boldsymbol{\mu}' = \boldsymbol{\mu}/(2^{13L} t_1)$ to yield a new system of equations

$$\begin{aligned} A_U^T \boldsymbol{\mu}' &= \boldsymbol{\epsilon}', \\ A_U \boldsymbol{\xi} - \mathbf{b}_U &= \boldsymbol{\epsilon}, \\ \mathbf{e}^T \boldsymbol{\mu}' &\geq 1, \\ \boldsymbol{\mu}' &\geq \mathbf{0}. \end{aligned}$$

Notice that $\boldsymbol{\epsilon}'$ and $\boldsymbol{\epsilon}$ are both bounded by 2^{-6L} in the ∞-norm. Therefore, by Lemma 3.10 there exists a pair of vectors $(\mathbf{x}'', \boldsymbol{\mu}'')$ to solve

$$\begin{aligned} A_U^T \boldsymbol{\mu}'' &= \mathbf{0}, \\ A_U \mathbf{x}'' - \mathbf{b}_U &= \mathbf{0}, \\ \mathbf{e}^T \boldsymbol{\mu}'' &\geq 1, \\ \boldsymbol{\mu}'' &\geq \mathbf{0}. \end{aligned}$$

Let $\mathbf{y} = \boldsymbol{\xi} - \mathbf{x}''$; then

$$A_U \mathbf{y} = \boldsymbol{\epsilon}$$

and $\boldsymbol{\epsilon}$ has strictly positive entries. If we take the inner product of both sides of the previous equation with $\boldsymbol{\mu}''$, then the left-hand side becomes zero. The right-hand side is the inner product of a strictly positive vector with $\boldsymbol{\mu}''$; this must be nonzero. This is a contradiction. ∎

Now we return to the thread of the previous argument. We know by definition that

$$\begin{aligned}\nu(\mathbf{x}^+ - \mathbf{x}_\sim, t^+) &= \|\Delta(\boldsymbol{\xi}^+)^{-1}A(\mathbf{x}^+ - \mathbf{x}_\sim)\|_2 \\ &\leq \|\Delta(\boldsymbol{\xi}^+)^{-1}\|_2 \cdot \|A\|_2 \cdot \|\mathbf{x}^+ - \mathbf{x}_\sim\|_2 \\ &\leq \frac{1}{t^+} \cdot \|t^+\Delta(\boldsymbol{\xi}^+)^{-1}\|_2 \cdot \|A\|_2 \cdot \|\mathbf{x}^+ - \mathbf{x}_\sim\|_2 \\ &\leq \frac{\max(t^+, 1)}{t^+} \cdot 2^{13L} \cdot 2^{L_0} \cdot \sqrt{n} \cdot 2^{-27L-10} \\ &\leq 2^{-10}.\end{aligned}$$

In the above calculation we used the fact that $1/t^+ \leq 2^{13L}$; this is true because the main loop of the algorithm ends before t drops below 2^{-13L}.

We have bounded two of the three intermediate steps from $\mathbf{x}_\sim$ to $\boldsymbol{\xi}_\sim$; the last bound we want to establish is

$$\nu(\boldsymbol{\xi}^+ - \boldsymbol{\xi}_\sim, t^+) \leq 1/1000.$$

To prove this, we return to the Renegar-Shub theorem. We apply the theorem with norm $\nu(\cdot, t^+)$, center of the ball at $\boldsymbol{\xi}^+$, and function f being $\nabla h(\cdot, t_\sim)$, which has its root at $\boldsymbol{\xi}_\sim$. We do not need a Newton starting point. Then we need to compute the parameters of the theorem. From Lemma 3.6 we conclude that β is at most $|t^+ - t_\sim| \cdot \sqrt{m}/t_\sim$, which is at most $1/1500$ by definition of $t_\sim$ Lemma 3.7 gives a bound of 1 on γ.

Then the theorem tells us that the distance from $\boldsymbol{\xi}^+$ to $\boldsymbol{\xi}_\sim$ is at most $1.5/1500$, i.e., $1/1000$.

We have shown that

$$\nu(\mathbf{x}^+ - \mathbf{x}_\sim, t^+) \leq 1/1000,$$

$$\nu(\boldsymbol{\xi}^+ - \boldsymbol{\xi}_\sim, t^+) \leq 1/1000,$$

and

$$\nu(\mathbf{x}^+ - \boldsymbol{\xi}^+, t^+) \leq 1/22;$$

thus

$$\nu(\mathbf{x}_\sim - \boldsymbol{\xi}_\sim, t^+) \leq 0.0475 \tag{3.29}$$

by the triangle inequality. This right-hand side is strictly less than $1/20$.

Now, finally, we apply Lemma 3.8 to switch bound (3.29) over to norm $\nu(\cdot, t_\sim)$. The relevant fact needed to apply the lemma is a bound on $\nu(\boldsymbol{\xi}^+ - \boldsymbol{\xi}_\sim, t^+)$ which was shown above to be $1/1000$. Therefore,

$$\nu(\mathbf{x}_\sim - \boldsymbol{\xi}_\sim, t_\sim) \leq \frac{1000}{999} \cdot 0.0475.$$

The right-hand side is bounded by $1/20$; this gives the result.

3.2.6. *Strongly polynomial time*

The results in the preceding subsections establish the main result of this chapter, namely, convex quadratic program can be solved in polynomial type. The structure of this proof was as follows. We first constructed an iterative interior point method that converges to a global minimum. After this algorithm runs for $O(\sqrt{m}L)$ iterations, we use a rounding argument to step to an exact global minimum. All known polynomial-time algorithms for convex quadratic programming and linear programming have the same structure. These algorithms are not strongly polynomial time because the number of arithmetic operations depends on the number of bits required to represent the inputs, and not just on the dimensions of the matrices in the input.

The simplex method for linear programming, invented by Dantzig (see Dantzig [1963] for an exposition), has an upper bound on the number of arithmetic steps that depends only on the dimensions of the matrices and not on the number of bits. The upper bound, however, is exponential in these dimensions, and hence does not establish any kind of polynomial-time result. Indeed, Klee and Minty [1972] showed that the exponential upper bound is attained by the simplex algorithm in the worst case.

It is an important open question whether convex quadratic programming or linear programming can be solved in strongly polynomial time. A closely related conjecture appears in Traub and Woźniakowski [1982]. They conjecture that there is no polynomial-time algorithm for linear programming in the real-number model. In the real-number model, the simplex algorithm requires an exponential number of steps, and the interior point method of this section does not terminate (there is no upper bound on how many steps are required before an exact global minimum can be obtained from the iterates). Note that this conjecture differs from the question of strongly polynomial time, since the underlying models of computation in the two questions are different (polynomial-time Turing machines versus real-number models).

In practice, the interior point method of this section would be run until the iterates differed from each other by no more than some small cutoff value. This would be taken as a machine approximation to the global minimum. The whole issue of rounding to an exact global minimum would not arise.

We conclude this section by briefly mentioning two results related strongly polynomial time algorithms for linear programming. First, Megiddo [1984] showed that linear programming can be solved in strongly polynomial time (linear time, in fact) if the number of variables is fixed—only the number of constraints may vary. This is useful primarily for geometric optimization problems, in which the variables of the problem are coordinates of a point in space.

A second related result is due to Tardos [1986], who showed that many linear programming problems that arise from combinatorial applications can be solved in strongly polynomial time.

Exercises

3.1 We will say a QP is in *standard form* if it has the following format:

$$\begin{aligned} \text{minimize} \quad & \tfrac{1}{2}\mathbf{x}'^T H'\mathbf{x}' + \mathbf{c}'^T\mathbf{x}' \\ \text{subject to} \quad & A'\mathbf{x}' = \mathbf{b}', \\ & \mathbf{x}' \geq \mathbf{0}. \end{aligned}$$

Show that a QP in the form given by (3.7) can always be transformed to the above standard form. Show that if H is positive semidefinite, then H' will also be positive semidefinite after your transformation.

3.2 Show how to modify the HKL algorithm to handle variables with missing lower or upper bounds (i.e., $l_i = -\infty$ or $u_i = \infty$).

3.3 (Brucker [1984]) Show how to modify the HKL algorithm so that the running time is $O(n)$ instead of $O(n \log n)$. [*Hint.* Instead of sorting the breakpoints and doing binary search, consider finding the *median* breakpoint b_m and checking whether $\lambda^* < b_m$, $\lambda^* = b_m$, or $\lambda^* > b_m$. After this test is made, the number of breakpoints can be cut in half, and then the problem can be solved recursively. Argue that the running time is $O(n)$.]

3.4 Under what circumstances does the set of points $\boldsymbol{\xi}(t)$ defined in this chapter converge as $t \rightarrow 0$?

3.5 Prove the claim that each term of (3.22) is at most 1 when $(\boldsymbol{\xi}_a, \boldsymbol{\mu}_a)$ is substituted for $(\mathbf{x}_b, \boldsymbol{\mu}_b)$.

3.6 For practical purposes it is important to know about the condition number of the matrix that is solved for a Newton method step. The condition number of a nonsingular matrix X is defined to be $\|X\|_2 \cdot \|X^{-1}\|_2$. Derive an upper bound on the condition number of

$$H + tA^T\Delta(\mathbf{x})^{-2}A$$

for the Renegar-Shub algorithm. Your bound should be of the form 2^{kL} for constant k that you must determine.

3.7 (Burton and Toint [1989]) The *inverse shortest paths* (ISP) problem is the following. You are given a directed graph G, a collection of *paths* $P_1, \ldots, P_s$ of G, and an initial cost vector $\mathbf{x}_0$ of nonnegative costs, one cost for each arc of G. (A *path* is a sequence of nodes joined by directed arcs such that no node is repeated.) Let the endpoints of P_i be u_i and v_i The problem is to find a nonnegative cost vector $\mathbf{x}$ with one entry per arc such that each of the specified paths P_i is the *shortest* path in G from u_i to v_i when the arc costs are $\mathbf{x}$. Moreover, among all such choices of $\mathbf{x}$, find the one satisfying these constraints that minimizes $\|\mathbf{x} - \mathbf{x}_0\|_2$.

Show that the ISP problem can be solved in polynomial time. It is suggested that you prove this by expressing an instance of ISP as a convex quadratic program with a polynomial number of constraints. [It may be necessary to introduce additional variables.]

3.8 In the next chapter we prove that if a (convex or nonconvex) quadratic program is feasible but unbounded, then there exists a feasible *ray*, that is, a set of points of the form $\{\mathbf{x}_r + t\mathbf{y}_r : t \geq 0\}$ for some vectors $\mathbf{x}_r, \mathbf{y}_r$ such that every point on the ray is feasible and such that the objective function tends to $-\infty$ as t tends to ∞.

Suppose that a convex quadratic program of the form (3.7) is unbounded. Derive a convex quadratic program based on $(H, \mathbf{c}, A, \mathbf{b})$ whose solution is a pair $\mathbf{x}_r, \mathbf{y}_r$ defined in the previous paragraph.

4

NONCONVEX QUADRATIC PROGRAMMING

In the last chapter, we derived a polynomial-time algorithm for convex quadratic programming. In the case of nonconvex programming, there is no known efficient general-purpose algorithm for global optimization because, as demonstrated in Chapter 2, this problem is **NP**-hard. In the first section of this chapter, we prove that the decision version of quadratic programming is actually **NP**-complete. This is mainly of theoretical interest—showing that the problem is **NP**-complete determines its complexity exactly.

In the second section, we look at some special cases of nonconvex QP that are also **NP**-hard. Even very simple forms of nonconvex QP are apparently intractable. The following section examines Ye's and Karmarkar's algorithm for an interesting class of nonconvex QP-type problems. This algorithm is polynomial time. Finally, we examine a simple exponential-time algorithm for the general case.

This chapter is of interest to practicing optimizers for the following reason. Often when faced with an unfamiliar type of optimization problem, an algorithm designer will break up the domain and model the objective function into simpler model problems. These simpler domains would include, for example, simplices, hypercubes or spheres. The simplified objective functions are often quadratic objective functions. One important point of this chapter is that some model problems are more difficult than others. For example, in this chapter we show that solving the model problem of (globally) minimizing a quadratic function over a simplex is **NP**-hard, whereas minimizing a quadratic function over a sphere can be solved in polynomial time.

4.1. QP is NP-complete

In this section we prove that nonconvex quadratic programming is actually in **NP**. Since we have already exhibited a polynomial transformation from SAT to nonconvex QP (proof of Theorem 2.5), the result of this section will prove that nonconvex QP is **NP**-complete (it has been known since Sahni [1974] that the problem is **NP**-hard). The result of this section was first proved by Vavasis [1990a], although the main ideas were around earlier (see, for example, Murty [1988]).

The importance of showing that a problem is **NP**-complete rather than merely **NP**-hard is primarily theoretical. The result indicates that QP probably does not lie in a more difficult complexity class such as the class of PSPACE-complete problems (see Garey and Johnson [1979]).

Recall that only decision problems can qualify for membership in **NP**. Accordingly, we state nonconvex quadratic programming as the following decision problem. Given a symmetric $n \times n$ matrix H, an $m \times n$ matrix A, an m-vector $\mathbf{b}$, an n-vector $\mathbf{c}$, and a number ζ, does there exist a real vector $\mathbf{x}$ such that $\frac{1}{2}\mathbf{x}^T H\mathbf{x}+\mathbf{c}^T\mathbf{x} \leq \zeta$ and $A\mathbf{x} \geq \mathbf{b}$? Here we assume that $(H, \mathbf{c}, A, \mathbf{b})$ have integer entries and ζ is rational.

A decision problem is in **NP** if there is a polynomial-length certificate for all **yes**-instances. The natural certificate to use for a **yes**-instance of the problem in the last paragraph is the vector $\mathbf{x}$ that satisfies $\frac{1}{2}\mathbf{x}^T H\mathbf{x}+\mathbf{c}^T\mathbf{x} \leq \zeta$, $A\mathbf{x} \geq \mathbf{b}$. Accordingly, the main thing we need to prove in this section is that if such an $\mathbf{x}$ exists, then such an $\mathbf{x}$ with a polynomial number of digits exists. (See Exercise 4.1 for an example of an optimization problem for which this approach towards proving membership in **NP** does not work.)

First, consider the case that the objective function $\frac{1}{2}\mathbf{x}^T H\mathbf{x} + \mathbf{c}^T\mathbf{x}$ has a global minimum on the set $A\mathbf{x} \geq \mathbf{b}$. Then, as demonstrated in the last chapter, there exists a global minimum $\mathbf{x}^*$ with at most a polynomial number of digits (in fact, the number of digits was linear in the problem size). This $\mathbf{x}^*$ can always be a certificate for a **yes**-instance. In other words, if there exists a real vector $\mathbf{x}$ such that $A\mathbf{x} \geq \mathbf{b}$, $\frac{1}{2}\mathbf{x}^T H\mathbf{x} + \mathbf{c}^T\mathbf{x} \leq \zeta$, then certainly $\mathbf{x}^*$ also satisfies these constraints.

Accordingly, the only remaining fact to prove is that such an $\mathbf{x}$ exists in the case that the objective function has no global minimum. There are two subcases here; either (a) the objective function is bounded from below but fails to achieve its minimum or (b) the objective function is unbounded. In fact, it turns out that case (a) is impossible for a quadratic objective function subject to a finite number of linear constraints, but we cannot rule out case (a) at this point of the argument.

In either case, the feasible region must be noncompact since we know that a global minimum of a continuous function always exists on a compact set. Since the set $A\mathbf{x} \geq \mathbf{b}$ is closed, then noncompactness implies unboundedness.

Let $f(\mathbf{x})$ denote the objective function $\frac{1}{2}\mathbf{x}^T H\mathbf{x} + \mathbf{c}^T\mathbf{x}$. Since the global minimum does not exist, there is a sequence of points $\mathbf{x}^{(0)}, \mathbf{x}^{(1)}, \ldots$ such that $f(\mathbf{x}^{(k)})$ either diverges to $-\infty$ or converges to the greatest lower bound of the values of $f(\mathbf{x})$ on the domain.

For each k define $v_k = \|\mathbf{x}^{(k)}\|_\infty$. The sequence $v_1, v_2, \ldots$ cannot have an upper bound; if it did, then the sequence $\mathbf{x}^{(k)}$ would lie in a compact set, contradicting our assumption that the minimum fails to exist. Now consider the sequence of quadratic programming problems QP_k given by

$$\begin{aligned}\text{minimize}\quad & \tfrac{1}{2}\mathbf{x}^T H\mathbf{x} + \mathbf{c}^T\mathbf{x}\\ \text{subject to}\quad & A\mathbf{x} \geq \mathbf{b},\\ & x_i \leq v_k, \text{ for } i = 1,\ldots,n,\\ & x_i \geq -v_k, \text{ for } i = 1,\ldots,n.\end{aligned}$$

Notice that $\mathbf{x}^{(k)}$ is feasible for this problem.

We redefine $\mathbf{x}^{(k)}$ now to be the global minimum of QP_k. Such a minimum exists since the feasible region of QP_k is nonempty and compact. Moreover, we know that under this new definition $f(\mathbf{x}^{(k)})$ still either tends to the greatest lower bound or diverges to $-\infty$. By passing to a subsequence, we can assume without loss of generality that $f(\mathbf{x}^{(k)})$ is strictly decreasing.

Suppose there is more than one global minimum of QP_k. Among all choices we select the minimum as in the previous chapter. Specifically, we select the minimum with the largest number of active inequalities. If there is still more than one minimum, we select the one with the largest number of zero coordinates. Further ties are broken arbitrarily.

Note that QP_k has exactly $m + 2n$ inequality constraints; assume they are numbered in some canonical order (say the order listed above). Let the *signature* of a particular $\mathbf{x}^{(k)}$ be the list indicating which of the $m + 2n$ inequalities are active and which coordinates are equal to zero. There are at most 2^{m+3n} possible signatures that a point could have. Since there are only a finite number, there must be at least one signature that occurs infinitely often in the sequence $\mathbf{x}^{(k)}$. By passing to a subsequence, we can assume (without loss of generality) that all the points in the sequence have the same signature.

As in Section 3.2, we construct a sequence of affine spaces Y_k based on the active inequalities. As before,

$$Y_k = \{x \in \mathbb{R}^n : M\mathbf{x} = \mathbf{b}_k\}.$$

Notice that the coefficient matrix M for Y_k will be independent of k; this follows from the assumption that all the points have the same signature. The right-hand-side $\mathbf{b}_k$ depends on k; in particular, its entries will either be coefficients from the original quadratic program or they will be v_k or $-v_k$. This is because some of the constraints $x_i \leq v_k$ or $x_i \geq -v_k$ will be active.

Applying the technique of the previous chapter, we conclude that the vector $\mathbf{x}^{(k)}$ is actually the solution to the nonsingular system of equations

$$\begin{aligned}H\mathbf{x}^{(k)} + \mathbf{c} &= M^T\boldsymbol{\mu}^{(k)},\\ M\mathbf{x}^{(k)} &= \mathbf{b}_k.\end{aligned}$$

The solution $(\mathbf{x}^{(k)}, \boldsymbol{\mu}^{(k)})$ to this system will depend linearly on $\mathbf{b}_k$, and hence linearly on v_k. Thus, $\mathbf{x}^{(k)} = \mathbf{x}_0 + v_k\mathbf{x}_1$ where $\mathbf{x}_0$ and $\mathbf{x}_1$ depend only on the original coefficients. Moreover, $\mathbf{x}_0$ and $\mathbf{x}_1$ will have a polynomial number of digits. In particular, $\mathbf{x}_0$ is the solution to the previous system with 0's substituted for v_k, and $\mathbf{x}_1$ is the solution to the previous system with 1's substituted for v_k and zeros substituted for all entries of $\mathbf{c}$ and for the rest of $\mathbf{b}$.

Finally, substituting the formula for $\mathbf{x}^{(k)}$ into the objective function, we arrive at the conclusion that $f(\mathbf{x}^{(k)}) = q(v_k)$ where $q(v_k)$ is a real-valued quadratic function of one variable whose coefficients require polynomial space. Since this holds for v_k arbitrarily large, we can define a ray $\mathbf{x}(v) = \mathbf{x}_0 + v\mathbf{x}_1$ which is feasible for the original QP for all v large enough and which has a decreasing objective function.

How large must v be in order for $\mathbf{x}(v)$ to be feasible for the original QP? Let v^* be the minimum value of v such that $\mathbf{x}(v)$ is feasible on $[v^*, +\infty)$, or zero, whichever is larger. If v^* is nonzero, then there must be an i that $\mathbf{a}_i^T\mathbf{x}(v^*) = b_i$ and such that $\mathbf{a}_i^T\mathbf{x}(v) < b_i$ for $v < v^*$. Therefore, the equation $\mathbf{a}_i^T\mathbf{x}(v^*) = b_i$ is a nondegenerate linear equation of one variable. Solving for v^* shows that it can be expressed as a rational number whose number of digits is polynomial in the length of the input. Therefore, we can translate the parameterization of $\mathbf{x}(v)$ so that, without loss of generality, $\mathbf{x}(v)$ is feasible for all $v \geq 0$.

Now finally we have reduced the original problem of showing the existence of a polynomial-length $\mathbf{x}$ to the following problem. Given a decreasing real-valued quadratic function $q(v)$ whose coefficients are rational and given a rational number ζ, find a point $v^* \geq 0$ such that $q(v^*) \leq \zeta$.

This is a simple problem to solve; we now indicate the argument. The only trick is to show that we need not take a square root. If the leading coefficient of q is zero, then q is linear and we simply solve the linear equation. If the leading coefficient of q is nonzero, then we write $q(v)$ as $q(v) = a(v - v_0)^2 + q_0$. This requires only rational operations. Notice that a must be negative in this case, since q is decreasing. Then the problem is to solve $(v - v_0)^2 \geq (\zeta - q_0)/a$. Now, if $(\zeta - q_0)/a \leq 1$, then we can take $v^* = \max(v_0 + 1, 0)$. If $(\zeta - q_0)/a \geq 1$, then we take $v^* = \max(v_0 + (\zeta - q_0)/a, 0)$. In all cases, we find a rational number v^* with a polynomial number of digits such that $v^* \geq 0$ and $q(v^*) \leq \zeta$.

Once this v^* is known, we can determine the certificate $\mathbf{x}$ for the original problem, namely $\mathbf{x} = \mathbf{x}(v^*)$. This requires only a polynomial number of digits.

This concludes the proof that quadratic programming is in **NP**.

THEOREM 4.1. *Let QPL be the following problem. Given $(H, \mathbf{c}, A, \mathbf{b})$, where H is an $n \times n$ matrix, $\mathbf{c}$ is an n-vector, A is an $m \times n$ matrix, and $\mathbf{b}$ is an m-vector, all with integer entries, and given a rational number ζ, does there exist an $\mathbf{x} \in \mathbb{R}^n$ such that $\frac{1}{2}\mathbf{x}^T H\mathbf{x} + \mathbf{c}^T\mathbf{x} \leq \zeta$ and $A\mathbf{x} \geq \mathbf{b}$? Then problem QPL is* **NP**-*complete.*

PROOF. Sahni's reduction or the reduction in Chapter 2 shows that any problem in **NP** can be polynomially transformed to QPL. In this section we have argued that there is a polynomial-length certificate for **yes**-instances of QPL, showing that the problem is in **NP**. ∎

Notice that as a byproduct of this proof we have shown the following theorem. There seems to be no way to prove this theorem simpler than the machinery of this section.

THEOREM 4.2. *Consider the problem of minimizing a quadratic objective function $f(\mathbf{x})$ subject to $A\mathbf{x} \geq \mathbf{b}$. Then the possible outcomes are: (a) The constraints are infeasible, (b) A global minimum exists, or (c) The objective function is unbounded from below in the feasible region. In particular, the following case is not possible: The objective function is bounded below but does not achieve its minimum.*

PROOF. The preceding argument shows that for a feasible QP that does not attain its minimum, there must exist a ray such that every point on the ray is feasible, and the objective function is a decreasing quadratic function on the ray. Such a function must tend to $-\infty$. ∎

4.2. Special cases of nonconvex QP

In this section we will list some special cases of nonconvex QP that are known to be **NP**-hard. Since the general case of QP was proved to lie in **NP** when expressed as a decision problem, all of the examples of this section are also **NP**-complete when posed as decision problems.

QP with simple bounds

This is the problem of minimizing $\frac{1}{2}\mathbf{x}^T H\mathbf{x} + \mathbf{c}^T\mathbf{x}$ subject to constraints $l_i \leq x_i \leq u_i$ for $i = 1, \ldots, n$. These constraints are sometimes called "box constraints." This problem is **NP**-hard by polynomial time transformation from SAT similar to the transformation proposed in the proof of Theorem 2.5. In that transformation we required simple bounds on the variables, namely, $0 \leq y_i \leq 1$. We also included constraints of the form $\mathbf{a}_i^T\mathbf{y} \geq b_i$ to capture the ith clause. For a setting of the variables corresponding to an assignment that satisfied the clause, we note from the reduction that $\mathbf{a}_i^T\mathbf{y} = b_i + v$ where v is a nonnegative integer between 0 and 2. We replace this constraint with an additional term in the objective function. The additional term takes the form $(\mathbf{a}_i^T\mathbf{y} - b_i - v_i)^2$ where v_i is a new variable bound to lie between 0 and 2.

This new form of QP has only simple-bound constraints on the variables. It still has the property that the objective function is nonnegative on the feasible region. One can prove that if the original boolean formula is satisfiable, then there is a setting of the variables in the quadratic program in which the value of the objective function is zero. One can also prove conversely that setting of the variables to make the objective function zero must be an assignment to the boolean variables that satisfies all clauses.

Quadratic knapsack problem

The quadratic knapsack problem (QKP) was introduced in Section 3.1. This problem is **NP**-hard if the matrix D is not positive semidefinite, i.e., if it has negative entries on its diagonal.

We transform the subset-sum problem to QKP. The subset sum problem was described in Chapter 2 and is **NP**-complete (this was Theorem 2.4). In this decision problem, the input is a list of n nonnegative integers $a_1, \ldots, a_n$ and an integer γ. The question is whether there is a subset $J \subset \{1, \ldots, n\}$ such that

$$\sum_{j \in J} a_j = \gamma.$$

This is expressed as the following quadratic knapsack problem:

$$\begin{aligned} \text{minimize} \quad & x_1(1 - x_1) + \cdots + x_n(1 - x_n) \\ \text{subject to} \quad & a_1 x_1 + \cdots + a_n x_n = \gamma, \\ & 0 \le x_i \le 1. \end{aligned} \tag{4.1}$$

Note that the objective function is nonnegative on the feasible region. Note also that the global minimum of the objective function is zero if and only if each x_i is set to 0 or 1, i.e., if and only if γ can be expressed as a sum of some subset of the integers $a_1, \ldots, a_n$.

Problems on a simplex

The *standard n-dimensional simplex* is defined to be the following subset of $\mathbb{R}^{n+1}$:

$$\Delta^n = \{(x_1, \ldots, x_{n+1}) : x_1 + \cdots + x_{n+1} = 1; x_i \ge 0 \text{ for } i = 1, \ldots, n+1\}.$$

Fig. 4.1 illustrates Δ^1 and Δ^2. This region is specified by one equation constraint and $n+1$ inequalities. It has exactly $n+1$ vertices.

The problem of minimizing $\frac{1}{2}\mathbf{x}^T H \mathbf{x} + \mathbf{c}^T \mathbf{x}$ subject to $\mathbf{x} \in \Delta^{n-1}$ is **NP**-hard. This fact was observed by Pardalos, Ye and Han [1989], and the proof follows a theorem by Motzkin and Straus [1965]. We first recall that the following problem is **NP**-hard: Given a graph G, find the largest clique contained as a subgraph in G. A *clique* is a graph with all pairs of vertices interconnected. A clique with m vertices is denoted as K_m. The **NP**-hardness of this problem follows from Theorem 2.3.

We can transform the clique problem to QP on a simplex. The transformation is as follows. Let G be an undirected graph with n vertices numbered

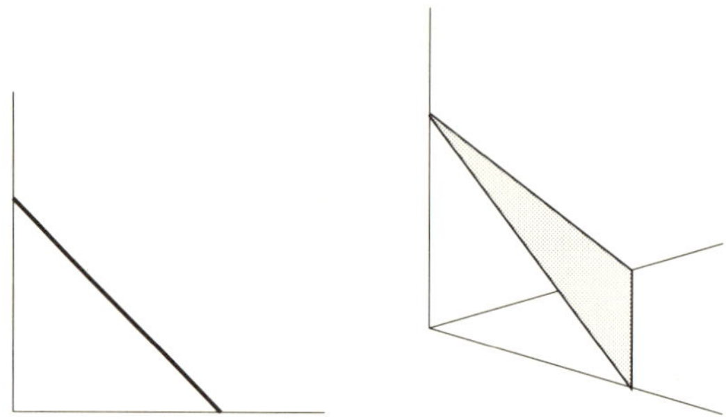

FIG. 4.1. One and two dimensional simplices.

$1, \ldots, n$. There will be a variable x_i for each vertex i. Let $E(G)$ be the edges of G, that is, a set of unordered pairs of the form (i, j) with $1 \leq i, j \leq n$. Let f be the following quadratic function from $\mathbb{R}^n$ to $\mathbb{R}$.

$$f(\mathbf{x}) = - \sum_{(i,j) \in E(G)} x_i x_j \tag{4.2}$$

(only one of the terms $x_i x_j$ and $x_j x_i$ occurs in the sum for each edge $(i, j) \in E(G)$.) We claim that the minimum value attained by $f(\mathbf{x})$ is

$$\frac{1}{2k} - \frac{1}{2}, \tag{4.3}$$

where k is the size of the maximum clique in G.

First, suppose there is a clique of size k. Then we can set the variables corresponding to clique vertices to $1/k$. The remaining variables are zero. Clearly this point is feasible. The objective function in this case will have $k(k-1)/2$ terms each of value $-1/k^2$, so the total objective function value is $-k(k-1)/(2k^2)$ which is $1/(2k) - 1/2$.

Now, we prove that the objective function cannot be lower than p where $p = 1/(2k) - 1/2$ and k is the size of the largest clique of G.

We state this as a lemma, which is due to Motzkin and Straus.

LEMMA 4.1. *The optimum value of f defined by* (4.2) *on* Δ^{n-1} *is given by* (4.3).

PROOF. We have already shown that the minimurn is no more than the value given by (4.3).

The proof for the other direction is by induction on n, the number of vertices of G. First, if $n = 2$ and G has one edge, then the minimum objective function value is $-1/4$. If G has two vertices and no edges, then the objective function is identically zero and the maximum clique size is 1.

Assume that $n > 2$ and examine the optimum value $\mathbf{x}^*$ of the QP. Let k^* be the size of the largest clique of G. We want to prove that $f(\mathbf{x}^*) \geq 1/(2k^*) - 1/2$. There are three cases.

Case 1. *For some* i, $x_i^* = 0$. In this case we can delete vertex i from G to yield a graph G' with $n - 1$ vertices. Note that the objective function f' for this graph G' also attains value $f(\mathbf{x}^*)$ (delete the ith entry from $\mathbf{x}^*$). Let $\mathbf{x}'$ be the optimum for f'; then we have just argued that $f(\mathbf{x}^*) \geq f'(\mathbf{x}')$. By the induction hypothesis, $f'(\mathbf{x}') = 1/(2k') - 1/2$, where k' is the size of the largest clique of G'. But G also contains this same clique, so $k^* \geq k'$. Thus we get the chain of inequalities

$$f(\mathbf{x}^*) \geq f'(\mathbf{x}') = \frac{1}{2k'} - \frac{1}{2} \geq \frac{1}{2k^*} - \frac{1}{2}$$

which proves the claim.

Case 2. *For all* i, $x_i^* > 0$, *and* $G \neq K_n$. In this case, we look at the KKT conditions at optimum. The only active constraint is the constraint $\mathbf{e}^T\mathbf{x} = 1$ where $\mathbf{e}$ is the vector of all 1's. This means that $\nabla f(\mathbf{x}^*) = \lambda\mathbf{e}$ for some value of λ, hence the entries of $\nabla f(\mathbf{x}^*)$ are equal to one another. Since we are assuming that $G \neq K_n$, at least one possible edge is absent. Assume without loss of generality that edge $(1, 2)$ is not present in G. Then we examine the first two entries of $\nabla f(\mathbf{x}^*)$ which are,

$$\frac{\partial f}{\partial x_1}(\mathbf{x}^*) = - \sum_{(1,j)\in E(G)} x_j^*$$

and

$$\frac{\partial f}{\partial x_2}(\mathbf{x}^*) = - \sum_{(2,j)\in E(G)} x_j^*.$$

Since these are equal, we conclude that $f(x_1^*+t, x_2^*-t, x_3^*, \ldots, x_n^*)$ is equal to $f(\mathbf{x}^*)$ for all choices of t (since there is no cross-term x_1x_2, the dependence on t is linear). This means that there is some value of t such that either x_1 or x_2 can be driven to zero and the value of the objective function is preserved. This reduces the problem to Case 1.

Case 3. $G = K_n$. In this case, we can compute analytically the minimum value of the objective function. Notice that the objective function in this case is equal to

$$\frac{(x_1^2 + \cdots + x_n^2) - (x_1 + \cdots + x_n)^2}{2}.$$

The second parenthesized term of the numerator is always 1 because of the constraint $\mathbf{e}^T\mathbf{x} = 1$. Therefore, the problem is to minimize $(\|\mathbf{x}\|_2^2 - 1)/2$. Using the KKT conditions, one can show that $\|\mathbf{x}\|_2$ is minimized on a simplex when each x_i is set to $1/n$. This proves the claim.

Thus, we have shown that the minimum of f is attained when each x_i for the vertices of a maximum clique of size k is set to $1/k$. ∎

4.3. Nonconvex quadratic functions on a sphere

It is tempting to conclude from the previous section that every nontrivial version of nonconvex quadratic programming is **NP**-hard. In fact, this is not the case; in this section we give an efficient algorithm for a certain interesting version of nonconvex quadratic minimization.

The problem we consider is

$$\begin{aligned} \text{minimize} \quad & \tfrac{1}{2}\mathbf{x}^T H\mathbf{x} + \mathbf{c}^T\mathbf{x} \\ \text{subject to} \quad & \mathbf{x}^T\mathbf{x} \leq 1. \end{aligned} \tag{4.4}$$

Here H is an $n \times n$ symmetric matrix and $\mathbf{c}$ is an n-vector. This is the problem of minimizing a quadratic objective function on an n-dimensional sphere. Strictly speaking, this is not quadratic programming at all since we have defined QP to include only linearly constrained problems. It is closely related enough, however, that it is worth discussing in this chapter.

A simple generalization of this problem is the minimization of a quadratic objective function over an ellipsoid. An ellipsoid is the set of points satisfying $(\mathbf{x}-\mathbf{c})^T B^{-1}(\mathbf{x}-\mathbf{c}) \leq 1$, where B is a positive definite symmetric matrix. The problem of minimizing a quadratic function over an ellipsoid can be transformed to (4.4) by a linear transformation, although the transformation may involve irrational numbers (the Cholesky factorization LL^T of B must be formed).

One reason that (4.4) is of interest is that it comes up as a subproblem in more general optimization algorithms. Specifically, there is a class of nonlinear optimization algorithms called *model trust region methods*. In these algorithms, a quadratic function is used as a "model" of the nonlinear objective function around the current iterate $\mathbf{x}_c$. The next step is to minimize the model function. In general, however, the model is expected to be accurate (the model is "trusted") only in a neighborhood of $\mathbf{x}_c$. Accordingly, the model quadratic function is minimized over a neighborhood of $\mathbf{x}_c$. If the 2-norm is used to define the neighborhood, a problem exactly of the form (4.4) is produced. See Fig. 4.2 for a 1-dimensional trust region step.

The model-trust region method is due to Levenberg [1944] and Marquardt [1963]. These authors considered only the case where H is positive definite. Moré [1978] proposed a robust algorithm with a convergence proof for the positive definite case. Gay [1979] and Sorenson [1982] proposed algorithms for the general case.

The solution to (4.4) will in general be irrational. For example, if H is negative definite and $\mathbf{c} = \mathbf{0}$ then the solution to (4.4) will be half the minimum eigenvalue of H, and the vector $\mathbf{x}$ will be the corresponding eigenvector. The smallest eigenvalue of a negative definite rational matrix will usually be irrational.

Accordingly, we cannot expect to exactly compute the minimum value of the objective function nor the optimum $\mathbf{x}$. The goal of this section is

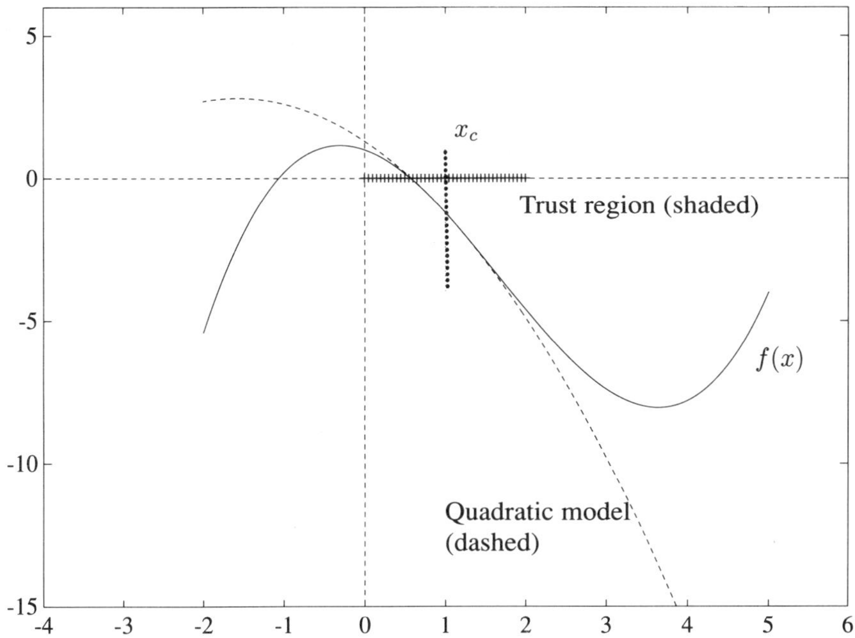

FIG. 4.2. Model trust region when $n = 1$.

therefore to show that $\mathbf{x}$ can be approximated efficiently. In particular, we show that p digits of $\mathbf{x}$ can be computed in time polynomial in p. This is analogous to the interior point results of Chapter 3, in which we showed that p digits of accuracy were produced in $O(p\sqrt{m})$ iterations. Alternatively, we can attempt to prove that the decision-problem version of (4.4) lies in **P**. We return to this topic in Exercise 4.6 through Exercise 4.8. The full proof that the decision problem lies in **P** is contained in Vavasis and Zippel [1990].

Observe that if the ∞-norm is used to define the trust region instead of the 2-norm, then one gets a quadratic objective function with simple bounds, which is an **NP**-hard problem as proved in the last section.

We now propose an algorithm for (4.4) due to Ye [1988]. A similar algorithm is due to Karmarkar [1989]. This algorithm in practice would be less efficient than the algorithms mentioned above, but has the advantage of leading to a simple polynomial time analysis. Ye [1990] proposes a Newton-method approach that would be competitive in practice, but we do not analyze it here.

The algorithm starts from the first and second-order necessary conditions for local minimality of (4.4). Before writing the conditions, we first dispense with the case that $\mathbf{x}^*$ is interior to the feasible region. In this case, $\mathbf{x}^*$ is a local minimum for an unconstrained quadratic function. This is possible only if $\mathbf{x}^*$ is the unconstrained global minimum of the objective function. Such a minimum exists only if H is positive semidefinite. If H is positive definite, then the global minimum is the root to $H\mathbf{x}^* = -\mathbf{c}$. Accordingly, this case is handled by (1) testing whether H is positive definite, and if so, (2) solving $H\mathbf{x}^* = -\mathbf{c}$, then (3) testing whether $\mathbf{x}^{*T}\mathbf{x}^* \leq 1$. If all these conditions hold, we have found the minimum. Solving a positive definite system of linear equations with integer coefficients can be done in polynomial time as proved in Chapter 2.

If H is positive semidefinite but not positive definite and has a minimum $\mathbf{x}^*$ interior to the unit sphere, then there must be a nonzero direction $\mathbf{d}$ from $\mathbf{x}^*$ such that $\mathbf{x}^* + t\mathbf{d}$ has the same value of the objective function as $\mathbf{x}^*$. This direction $\mathbf{d}$ is a eigenvector of H corresponding to the eigenvalue 0. Therefore, in this case, there is also a global minimum $\mathbf{x}^*$ satisfying $\mathbf{x}^{*T}\mathbf{x}^* = 1$.

Thus, we can focus attention on the case that $\mathbf{x}^{*T}\mathbf{x}^* = 1$. In this situation the KKT conditions are as follows. There exists μ^* such that

$$\begin{aligned} H\mathbf{x}^* + \mathbf{c} &= -\mu^*\mathbf{x}^*, \\ \mathbf{x}^{*T}\mathbf{x}^* &= 1, \\ \mu^* &\geq 0. \end{aligned} \tag{4.5}$$

The KKT conditions are necessary for local minimality of (4.4); this was the subject of Exercise 1.4. We now derive second-order necessary conditions for this problem. Let $\mathbf{v}_0$ be a vector such that $\mathbf{x}^{*T}\mathbf{v}_0 < 0$. Let

$$\mathbf{v} = \mathbf{v}_0 \cdot \frac{-2\mathbf{x}^{*T}\mathbf{v}_0}{\mathbf{v}_0^T\mathbf{v}_0}.$$

Then $\mathbf{v}$ is a positive scalar multiple of $\mathbf{v}_0$ satisfying $-2\mathbf{x}^{*T}\mathbf{v} = \mathbf{v}^T\mathbf{v}$. Note that

$$\begin{aligned} (\mathbf{x}^* + \mathbf{v})^T(\mathbf{x}^* + \mathbf{v}) &= \mathbf{x}^{*T}\mathbf{x}^* + 2\mathbf{x}^{*T}\mathbf{v} + \mathbf{v}^T\mathbf{v} \\ &= 1 + 2\mathbf{x}^{*T}\mathbf{v} - 2\mathbf{x}^{*T}\mathbf{v} \\ &= 1. \end{aligned}$$

Therefore, $\mathbf{x}^* + \mathbf{v}$ is feasible for (4.4). Since $\mathbf{x}^*$ is supposed to be a global (and local) minimum, the value of the objective function at $\mathbf{x}^* + \mathbf{v}$ must be no less than the value at $\mathbf{x}^*$. Multiplying out and simplifying, we are left with the inequality

$$\mathbf{v}^T H\mathbf{x}^* + \frac{1}{2}\mathbf{v}^T H\mathbf{v} + \mathbf{c}^T\mathbf{v} \geq 0.$$

We can write the first and last terms as $(H\mathbf{x}^* + \mathbf{c})^T\mathbf{v}$ and plug in the first equation of (4.5) to get

$$-\mu^*\mathbf{x}^{*T}\mathbf{v} + \frac{1}{2}\mathbf{v}^T H\mathbf{v} \geq 0.$$

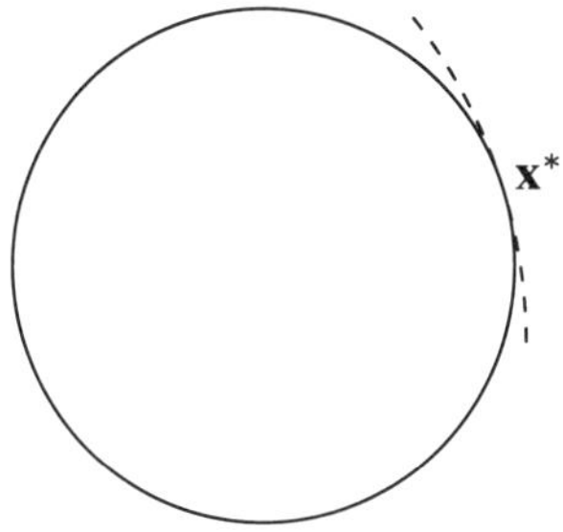

FIG. 4.3. A level curve at the optimum point in two dimensions.

Since $-2\mathbf{x}^{*T}\mathbf{v} = \mathbf{v}^T\mathbf{v}$, we can substitute to get

$$\frac{1}{2}\mu^*\mathbf{v}^T\mathbf{v} + \frac{1}{2}\mathbf{v}^T H\mathbf{v} \geq 0$$

which may be rewritten as

$$\mathbf{v}^T(H + \mu^* I)\mathbf{v} \geq 0. \tag{4.6}$$

This holds for $\mathbf{v}$, and hence also for $\mathbf{v}_0$. Therefore, we can conclude that (4.6) holds for all $\mathbf{v}$ such that $\mathbf{x}^{*T}\mathbf{v} < 0$. But then by changing the sign it holds for all $\mathbf{v}$ such that $\mathbf{x}^{*T}\mathbf{v} > 0$ as well. Finally, since the function $g(\mathbf{v}) = \mathbf{v}^T(H + \mu^* I)\mathbf{v}$ is continuous, it holds for all $\mathbf{v}$ such that $\mathbf{x}^{*T}\mathbf{v} = 0$ as well. Thus, (4.6) holds for all vectors $\mathbf{v}$. This is a condition that must be satisfied at a global minimum; this condition is equivalent to

$$H + \mu^* I \text{ is positive semidefinite.} \tag{4.7}$$

This condition implies, among other things, that the level curve of the objective function passing through $\mathbf{x}^*$ bends "outward" from the unit sphere. See Fig. 4.3 for an example in two dimensions. Note that this level curve is tangent to the unit sphere because of the KKT conditions.

It is a unique feature of the sphere-constrained quadratic problem that (4.5) and (4.7) are in fact sufficient for global minimality. In particular, it is not possible for (4.5) and (4.7) to be satisfied with differing values of μ or of the objective function. This is the content of the following two lemmas.

Let H be factored as $Q^T DQ$ where Q is an $n \times n$ orthogonal matrix and D is diagonal. Note that such a factorization always exists for a symmetric real matrix; see, for example, Golub and Van Loan [1989]. We will frequently refer to Q and D in the analysis that follows; they are not needed, however, for the algorithm itself. The diagonal entries of D are denoted as d_{ii}. These are the eigenvalues of H. Let μ_0 be the negative of the smallest d_{ii}, i.e., the negative of the smallest eigenvalue of H. Then clearly $H + \mu_0 I$ is positive semidefinite, and $H + \mu I$ is positive definite if and only if $\mu > \mu_0$.

LEMMA 4.2. *Let* $\mathbf{w}$ *be a nonzero vector, and suppose* μ_1 *and* μ_2 *are real numbers such that* $\mu_0 \leq \mu_1 < \mu_2$. *Suppose* $\mathbf{x}_1$ *is a solution to* $(H+\mu_1 I)\mathbf{x}_1 = \mathbf{w}$ *and* $\mathbf{x}_2$ *is a solution to* $(H + \mu_2 I)\mathbf{x}_2 = \mathbf{w}$. *Then*

$$\|\mathbf{x}_1\|_2 > \|\mathbf{x}_2\|_2.$$

PROOF. Since $\mathbf{w} \neq \mathbf{0}$, $\mathbf{x}_1$ and $\mathbf{x}_2$ are both nonzero. Since $\mu_2 > \mu_0$, the matrix $H + \mu_2 I$ is positive definite and hence invertible. We compute

$$\begin{aligned} Q\mathbf{x}_2 &= Q(H + \mu_2 I)^{-1}\mathbf{w} \\ &= Q(H + \mu_2 I)^{-1}(H + \mu_1 I)\mathbf{x}_1 \\ &= (D + \mu_2 I)^{-1}(D + \mu_1 I)Q\mathbf{x}_1. \end{aligned}$$

Since Q is orthogonal, $Q\mathbf{x}_2$ has the same norm as $\mathbf{x}_2$ and a similar statement holds for $\mathbf{x}_1$. Let $\mathbf{y}_1 = Q\mathbf{x}_1$ and $\mathbf{y}_2 = Q\mathbf{x}_2$ (both nonzero), and let y_{1i} and y_{2i} denote their entries. Then $\mathbf{y}_2$ is equal to the above diagonal matrix product applied to $\mathbf{y}_1$, hence

$$y_{2i} = \frac{d_{ii} + \mu_1}{d_{ii} + \mu_2} y_{1i}.$$

For each i the above fraction is nonnegative and less than 1 since $d_{ii} + \mu_1 \geq d_{ii} + \mu_0 \geq 0$ and $\mu_2 > \mu_1$. Therefore, the 2-norm of $\mathbf{y}_2$ is strictly less than the 2-norm of $\mathbf{y}_1$. ∎

LEMMA 4.3. *Suppose* $(\mathbf{x}_1, \mu_1)$ *and* $(\mathbf{x}_2, \mu_2)$ *are each pairs of solutions to* (4.5) *and* (4.7). *Then* $\mu_1 = \mu_2$ *and*

$$\frac{1}{2}\mathbf{x}_1^T H\mathbf{x}_1 + \mathbf{c}^T\mathbf{x}_1 = \frac{1}{2}\mathbf{x}_2^T H\mathbf{x}_2 + \mathbf{c}^T\mathbf{x}_2.$$

PROOF. We prove the first statement by contradiction. Suppose this statement is false; then, without loss of generality $\mu_1 < \mu_2$. By (4.7), $\mu_1 \geq \mu_0$. We notice that $\mathbf{x}_i$ for $i = 1, 2$ is a solution of $(H + \mu_i I)\mathbf{x}_i = -\mathbf{c}$. Moreover, $-\mathbf{c} \neq \mathbf{0}$ since the equation involving $\mathbf{x}_2$ is nonsingular, and we know that $\mathbf{x}_2 \neq \mathbf{0}$. Therefore, the previous lemma tells us that $\mathbf{x}_1$ and $\mathbf{x}_2$ have different 2-norms. This contradicts (4.5), where we required that they both have 2-norms equal to 1.

Thus, we have shown that $\mu_1 = \mu_2$. Call this number μ^*. With a long calculation, we now prove that the objective function values are equal. If we let $f(\mathbf{x})$ denote the objective function, then

$$\begin{aligned} f(\mathbf{x}_1) - f(\mathbf{x}_2) &= \frac{1}{2}\mathbf{x}_1^T H\mathbf{x}_1 + \mathbf{c}^T\mathbf{x}_1 - \frac{1}{2}\mathbf{x}_2^T H\mathbf{x}_2 - \mathbf{c}^T\mathbf{x}_2 \\ &= \frac{1}{2}\mathbf{x}_1^T H(\mathbf{x}_1 - \mathbf{x}_2) + \frac{1}{2}\mathbf{x}_2^T H(\mathbf{x}_1 - \mathbf{x}_2) + \mathbf{c}^T(\mathbf{x}_1 - \mathbf{x}_2) \end{aligned}$$

$$
\begin{aligned}
&= \frac{1}{2}(H\mathbf{x}_1 + \mathbf{c})^T(\mathbf{x}_1 - \mathbf{x}_2) + \frac{1}{2}(H\mathbf{x}_2 + \mathbf{c})^T(\mathbf{x}_1 - \mathbf{x}_2) \\
&= -\mu^*\mathbf{x}_1^T(\mathbf{x}_1 - \mathbf{x}_2) - \mu^*\mathbf{x}_2^T(\mathbf{x}_1 - \mathbf{x}_2) \\
&= -\mu^*(\mathbf{x}_1 + \mathbf{x}_2)^T(\mathbf{x}_1 - \mathbf{x}_2) \\
&= -\mu^*\left(\mathbf{x}_1^T\mathbf{x}_1 - \mathbf{x}_1^T\mathbf{x}_2 + \mathbf{x}_2^T\mathbf{x}_1 - \mathbf{x}_2^T\mathbf{x}_2\right) \\
&= 0. \qquad (4.8)
\end{aligned}
$$

∎

Therefore, the problem of solving (4.4) is reduced to the problem of finding μ^* and $\mathbf{x}^*$ to satisfy (4.5) and (4.7). We now prove that this pair always exists. (Actually, existence is already proved since (4.4) has a solution by compactness, and the two conditions are satisfied at this solution. The goal for now, however, is to come up with a more constructive proof of existence.) Define

$$p(\mu) = \|(H + \mu I)^{-1}\mathbf{c}\|_2$$

for $\mu > \mu_0$. Notice that p is a nonincreasing function (by Lemma 4.2) and is continuous. We want to show that it crosses through the value 1 for a $\mu \geq \max(0, \mu_0)$; this would establish the existence of a solution to (4.5) and (4.7). First, we show that it definitely achieves values smaller than 1. Define

$$m = \|H\|_\infty + \|\mathbf{c}\|_2 + 1;$$

then one checks using standard matrix norm estimates that $\|(H + mI)^{-1}\mathbf{c}\|_2$ is less than 1.

Now, we have to show that $p(\mu)$ is bigger than 1 for μ small enough. There are several cases to consider. First suppose that $\mu_0 \geq 0$. Then we investigate what happens to $p(\mu)$ as $\mu \downarrow \mu_0$. If we let $\mathbf{c}' = Q\mathbf{c}$, then we see that $p(\mu) = \|(D + \mu I)^{-1}\mathbf{c}'\|_2$, and μ is tending to the negative of the smallest diagonal entry of D. Therefore, we conclude that $p(\mu)$ tends to $+\infty$ in this situation—and, in particular, is greater than 1—unless $\mathbf{c}'$ has zeros in all the positions in which $D + \mu_0 I$ has zeros. The condition that $c_i' = 0$ for all i such that $d_{ii} + \mu_0 = 0$ is equivalent to the condition that $\mathbf{c}$ lies in the image of $H + \mu_0 I$. Accordingly, we have shown that $p(\mu)$ increases past 1 as long as $\mathbf{c}$ does not lie in the image of $H + \mu_0 I$.

The second case is that $\mu_0 \geq 0$ and $\mathbf{c}$ lies in the image of $H + \mu_0 I$. If $p(\mu)$ surpasses 1 for $\mu > \mu_0$ then we are done (we have proved it must cross through 1). Otherwise, suppose $p(\mu) \leq 1$ for all $\mu > \mu_0$. Then we can take $\mu^* = \mu_0$, and we take $\mathbf{x}_0$ to be the limit point of the sequence of vectors $-(H + \mu I)^{-1}\mathbf{c}$ as $\mu \downarrow \mu_0$. This limit point exists and lies in the unit ball because $p(\mu) \leq 1$. Moreover, it satisfies $(H + \mu_0 I)\mathbf{x}_0 = -\mathbf{c}$ because the mapping $(\mathbf{x}, \mu) \mapsto (H + \mu I)\mathbf{x}$ is continuous everywhere.

As noted above, we can now get a point $\mathbf{x}^*$ satisfying $\mathbf{x}^{*T}\mathbf{x}^* = 1$ (and therefore the conditions for global minimality) by adding to $\mathbf{x}_0$ a vector $\mathbf{w}$ in the null-space of $H + \mu_0 I$.

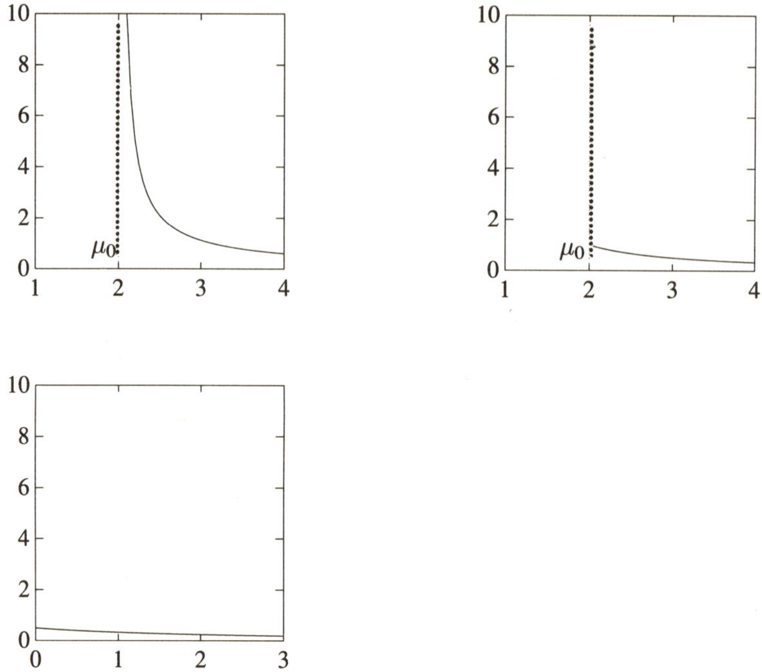

FIG. 4.4. The three cases for $p(\mu)$.

The third case is when $\mu_0 < 0$, so we consider $p(0)$. Suppose $p(0) < 1$; this means that $H^{-1}\mathbf{c}$ is contained in the interior of the unit ball. But then $-H^{-1}\mathbf{c}$ is the global minimum of the objective function when it is unconstrained, and hence we are in the case of a unique interior global minimum, which was settled earlier. The three cases are illustrated in Fig. 4.4. Thus, we have shown in all cases that $\mathbf{x}^*$ and μ^* exist to satisfy (4.5) and (4.7). Moreover, the existence argument suggests a bisection algorithm to find μ^*.

We start with μ^* known to lie in the interval $[0, m]$. We choose the midpoint of the interval, say μ_m, and perform the following tests. First, we check whether $\mu_m \geq \mu_0$. This test is equivalent to testing whether symmetric matrix $H + \mu_m I$ is positive semidefinite. This test is done via LDL^T factorization. If the matrix is positive semidefinite, then we know $\mu_m \geq \mu_0$.

If it turns out that $\mu_m < \mu_0$, then we know that μ_m is too small to be μ^* because (4.7) would fail to be satisfied.

If it turns out that $\mu_m = \mu_0$, this case can be handled with a complicated analysis. The simplest approach, however, is to reject this trial and pick μ_m to be a slight perturbation of μ_0.

Finally, if it turns out that $\mu_m > \mu_0$, then we solve the linear system

$$(H + \mu_m I)\mathbf{x} = -\mathbf{c}$$

to yield $\mathbf{x}$. Then we compare $\mathbf{x}^T\mathbf{x}$ to 1. If $\mathbf{x}^T\mathbf{x} > 1$ then by Lemma 4.2 $\mu^* > \mu_m$. If $\mathbf{x}^T\mathbf{x} \leq 1$ then $\mu^* < \mu_m$.

Thus, for any μ_m we can determine which side of μ^* it lies on; this is the main ingredient for a binary search to compute μ^*. The bracket containing μ^* is of width $2^{-k}m$ on the kth iteration.

Let the right endpoints of the intervals generated by the binary search be denoted as $\mu^{(k)}$. The preceding analysis shows that we can get a bound on the convergence rate of $\mu^{(k)}$ to μ^*. We would now like to establish a similar result for the convergence of $\mathbf{x}^{(k)}$ to $\mathbf{x}^*$.

In this regard, define

$$\mathbf{x}^{(k)} = -(H + \mu^{(k)}I)^{-1}\mathbf{c}.$$

How fast does $\mathbf{x}^{(k)}$ converge to the optimum? We can get a bound on the objective function using a calculation similar to (4.8).

We first treat the case that $\mu^* > \mu_0$, so that $p(\mu^*) = 1$. We have the following calculation:

$$\begin{aligned}
f(\mathbf{x}^{(k)}) - f(\mathbf{x}^*) &= \frac{1}{2}\mathbf{x}^{(k)^T}H\mathbf{x}^{(k)} + \mathbf{c}^T\mathbf{x}^{(k)} - \frac{1}{2}\mathbf{x}^{*T}H\mathbf{x}^* - \mathbf{c}^T\mathbf{x}^* \\
&= \frac{1}{2}\mathbf{x}^{(k)^T}H(\mathbf{x}^{(k)} - \mathbf{x}^*) + \frac{1}{2}\mathbf{x}^{*T}H(\mathbf{x}^{(k)} - \mathbf{x}^*) \\
&\qquad + \mathbf{c}^T(\mathbf{x}^{(k)} - \mathbf{x}^*) \\
&= \frac{1}{2}(H\mathbf{x}^{(k)} + \mathbf{c})^T(\mathbf{x}^{(k)} - \mathbf{x}^*) + \frac{1}{2}(H\mathbf{x}^* + \mathbf{c})^T(\mathbf{x}^{(k)} - \mathbf{x}^*) \\
&= -\frac{1}{2}\mu^{(k)}\mathbf{x}^{(k)^T}(\mathbf{x}^{(k)} - \mathbf{x}^*) - \frac{1}{2}\mu^*\mathbf{x}^{*T}(\mathbf{x}^{(k)} - \mathbf{x}^*) \\
&= \frac{1}{2}(\mu^* - \mu^{(k)})\mathbf{x}^{(k)^T}(\mathbf{x}^{(k)} - \mathbf{x}^*) \\
&\qquad - \frac{1}{2}\mu^*(\mathbf{x}^{(k)} + \mathbf{x}^*)^T(\mathbf{x}^{(k)} - \mathbf{x}^*) \\
&= \frac{1}{2}(\mu^* - \mu^{(k)})\mathbf{x}^{(k)^T}(\mathbf{x}^{(k)} - \mathbf{x}^*) \\
&\qquad - \frac{1}{2}\mu^*(\mathbf{x}^{(k)^T}\mathbf{x}^{(k)} - 1).
\end{aligned} \tag{4.9}$$

Now we can bound these two terms separately. We start with the first term; the vectors $\mathbf{x}^{(k)}$ and $\mathbf{x}^*$ have norm at most 1, and the quantity $(\mu^* - \mu_k)$ is at most $2^{-k}m$, hence this term is at most $2^{-k}m$.

The second term is equal to $-\frac{1}{2}\mu^*(p(\mu^{(k)})^2 - 1)$. The number $-\mu^*$ is at most m in magnitude. We know that $p(\mu^{(k)})^2 \in [0, 1]$, and moreover, $p(\mu^{(k)})^2$ tends to 1; the only question is how fast. Letting $\mathbf{c}'$ be as above, we notice that

$$Q\mathbf{x}^{(k)} = -(D + \mu^{(k)} I)^{-1}\mathbf{c}'$$

so we have

$$\begin{aligned} p(\mu^{(k)})^2 &= \sum_{i=1}^{n} \left(\frac{c_i'}{d_{ii} + \mu^{(k)}} \right)^2 \\ &= \sum_{i=1}^{n} \left(\frac{c_i'}{d_{ii} + \mu^*} \right)^2 \left(\frac{d_{ii} + \mu^*}{d_{ii} + \mu^{(k)}} \right)^2 . \end{aligned} \tag{4.10}$$

The summation on the right-hand side would be 1 if the second parenthesized factor were absent—this is because $p(\mu^*) = 1$. Let that factor be t_i^2, i.e.,

$$\begin{aligned} t_i &= \frac{d_{ii} + \mu^*}{d_{ii} + \mu^{(k)}} \\ &= 1 - \frac{\mu^{(k)} - \mu^*}{d_{ii} + \mu^{(k)}} \\ &\geq 1 - \frac{\mu^{(k)} - \mu^*}{\mu^{(k)} - \mu_0} \\ &\geq 1 - \frac{\mu^{(k)} - \mu^*}{\mu^* - \mu_0}. \end{aligned} \tag{4.11}$$

We want to prove that t_i tends to 1. Clearly $t_i \leq 1$, and the preceding inequality gives a lower bound on t_i. To make use of this inequality, we need an upper bound for $\mu^{(k)} - \mu^*$ (which has already been established), and we also need a lower bound for $\mu^* - \mu_0$.

Accordingly, we now derive a lower bound for $\mu^* - \mu_0$. Assume as in Chapter 3 that H and $\mathbf{c}$ have integer entries requiring a total of L bits to write. Our results up to now have assumed merely that H and $\mathbf{c}$ have real entries.

Notice that μ_0 is a solution to $\det(H + \mu I) = 0$, which is a degree-n polynomial equation in μ. All coefficients of the polynomial $\det(H + \mu I)$ have magnitude at most $n!2^{n+L}$; call this number s.

Similarly, μ^* is a solution to a polynomial equation. Specifically, we apply Cramer's rule to the equation $(H + \mu^* I)\mathbf{x}^* = -\mathbf{c}$ (recall that $H + \mu^* I$ is nonsingular since we are treating the case that $\mu^* > \mu_0$). Let $H_i(\mu)$ be the matrix (depending on μ) that results from replacing the ith column of $H + \mu I$ with $-\mathbf{c}$. Then μ^* is a root to

$$\sum_{i=1}^{n} \left(\frac{\det(H_i(\mu))}{\det(H + \mu I)} \right)^2 = 1.$$

Multiplying both sides by $\det(H + \mu I)^2$ gives a polynomial of degree $2n$ in μ. Each coefficient of this polynomial has magnitude at most $(n!2^{n+L})^2$; call this number t.

We state the main result on the separation polynomial roots and derive bounds on the convergence rate of the algorithm. The separation result itself is proved in an appendix to this section. For this theorem, $\mathbb{Z}[x]$ denotes the ring of polynomials with x as indeterminate and coefficients in $\mathbb{Z}$, the set of integers. If $p(x)$ is a polynomial in $\mathbb{R}[x]$, then $\|p\|$ denotes the maximum of the absolute value of the coefficients of p.

THEOREM 4.3. *Let α be a root of a nonzero polynomial $f(x) \in \mathbb{Z}[x]$, and let β be a root of a nonzero polynomial $g(x) \in \mathbb{Z}[x]$. Suppose $\alpha \neq \beta$, and suppose $f(x), g(x)$ have degree d, e respectively. Then*

$$|\alpha - \beta| \geq 2^{-2de-1} \|f\|^{-e} \|g\|^{-d}.$$

Now we apply Theorem 4.3 to conclude $\mu^* - \mu_0$ is bounded below by $2^{-4n^2-1} s^{-2n} t^{-n}$, which is at least 2^{-q} where $q = O(n^2 \log n + nL)$. Since $\mu^{(k)} - \mu^*$ at most $2^{-k}m$, a lower bound on t_i in (4.11) is $1 - 2^{q-k}m$, so a lower bound on t_i^2 would be $1 - 2^{q-k+1}m$. This is therefore a lower bound on (4.10). Thus, the second term of (4.9) differs from zero by at most $2^{q-k+1}m^2$.

Accordingly, the value of the objective function converges linearly, and the kth iteration is at most $2^{q'-k}$ above optimal, where $q' = O(n^2 \log n + nL)$.

So far we have been looking at the case that $\mu^* > \mu_0$. We next consider the case that $\mu^* = \mu_0$. This case can be detected after $O(n^2 \log n + nL)$ iterations, because $\mu^{(k)}$ will come too close to μ_0 in this situation (i.e., closer than the distance permitted by Theorem 4.3).

Recall that if $\mu^* = \mu_0$ then $\mathbf{c}$ lies in the image of the singular matrix $H + \mu_0 I$. We consider how the sequence $\mathbf{x}^{(k)}$ defined above behaves in this situation. By examining the transformed vector $Q\mathbf{x}^{(k)}$, we can conclude that $\mathbf{x}^{(k)}$ must converge to a solution of $(H + \mu_0 I)\mathbf{x} = -\mathbf{c}$ that satisfies $\mathbf{x}^T\mathbf{x} \leq 1$.

To solve (4.4) we must find a vector in the null-space of $H + \mu_0 I$; this must be accomplished without knowing μ_0 exactly. We now show how to compute a vector $\mathbf{y}$ that is an approximation a null-space vector.

Let $\mathbf{b}$ be a unit vector *not* in the image of $H + \mu_0 I$. To come up with such a vector $\mathbf{b}$, let $\mathbf{q}$ denote the ith column of Q^T, where i is an index such that $d_{ii} = -\mu_0$. Note that this $\mathbf{q}$ is an eigenvector of H. Suppose $\mathbf{q}^T\mathbf{b} \neq 0$. Then $(H + \mu_0 I)\mathbf{x} = \mathbf{b}$ has no solution (multiply both sides by $\mathbf{q}^T$; the left side is zero). Thus, $\mathbf{b}$ is not in the image of $H + \mu_0 I$.

Accordingly, we need a unit vector $\mathbf{b}$ not orthogonal to $\mathbf{q}$. We have not computed $\mathbf{q}$, but we know that among $\mathbf{e}_1, \ldots, \mathbf{e}_n$ (the columns of the identity matrix), at least one of these vectors will not be orthogonal to $\mathbf{q}$. Moreover, at least one of $\mathbf{e}_1, \ldots, \mathbf{e}_n$ will have an inner product of absolute value at least $1/\sqrt{n}$ with $\mathbf{q}$. Therefore, in the material that follows, we can assume that each $\mathbf{e}_i$ is substituted for $\mathbf{b}$; at least one will work. Indeed, a randomly chosen unit-length vector $\mathbf{b}$ will probably do. The expected value of $|\mathbf{q}^T\mathbf{b}|$ is about $2/(\pi\sqrt{n})$ for large n for $\mathbf{b}$ chosen uniformly at random (see Kuczyński and Woźniakowski [1990] for results in this direction).

Thus, assume we have a unit-length vector $\mathbf{b}$ such that $|\mathbf{q}^T\mathbf{b}| \geq 1/\sqrt{n}$.

Next, we compute the solution to the system $(H + \mu_1 I)\mathbf{y} = \mathbf{b}$, where μ_1 is greater than μ_0 but very close to it (see below). If we let $\mathbf{z} = Q\mathbf{y}$ and $\mathbf{b}' = Q\mathbf{b}$ then we see that $(D + \mu_1 I)\mathbf{z} = \mathbf{b}'$, i.e.,

$$z_j = \frac{b'_j}{d_{jj} + \mu_1}$$

for $j = 1, \ldots, n$. If we let i be the index of the distinguished column of Q^T selected above, then we see that b'_i is exactly equal to $\mathbf{q}^T\mathbf{b}$. Thus, $z_i = b'_i/(\mu_1 - \mu_0) \geq 1/[\sqrt{n}(\mu_1 - \mu_0)]$. Thus, $\mathbf{z}$ has 2-norm at least $1/[\sqrt{n}(\mu_1 - \mu_0)]$. This means that $\mathbf{y}$ has the same lower bound on its 2-norm. Thus, $\mathbf{y}$ will be a large vector since $\mu_1 - \mu_0$ is small.

Next, assume $\mathbf{x}$ is the solution to $(H + \mu_1 I)\mathbf{x} = -\mathbf{c}$. As noted above, $\mathbf{x}$ lies within the unit sphere. Then compute a vector $\mathbf{w} = \mathbf{x} + \alpha\mathbf{y}$ such that $\mathbf{w}^T\mathbf{w} = 1 - \epsilon$, where $\epsilon \geq 0$ is small. Note that this is possible because $\mathbf{x}^T\mathbf{x} \leq 1$ and $\mathbf{y} \neq \mathbf{0}$. The scalar α is computed as a rational approximate solution to the one-variable quadratic equation $(\mathbf{x} + \alpha\mathbf{y})^T(\mathbf{x} + \alpha\mathbf{y}) = 1$. (We would ideally want the exact solution, but the exact solution is probably irrational.) Note that by the triangle inequality, $\|\alpha\mathbf{y}\| \leq 2 - \epsilon$, hence $|\alpha| \leq 2/\|\mathbf{y}\|$. This implies that

$$|\alpha| \leq 2(\mu_1 - \mu_0)\sqrt{n}.$$

Note that $H\mathbf{w} + \mu_1\mathbf{w} = -\mathbf{c} + \alpha\mathbf{b}$ by choice of $\mathbf{w}$.

The claim is that $\mathbf{w}$ is an approximation to $\mathbf{x}^*$ because it is the sum of a "near" solution to $(H + \mu_0 I)\mathbf{x} = -\mathbf{c}$ and a vector $\alpha\mathbf{y}$ "nearly" in the null-space of $H + \mu_0 I$. Now we do a calculation to determine how far $\mathbf{w}$ is from the optimal $\mathbf{x}^*$ in terms of the objective function. The second line is derived from the first using the same intermediate steps as (4.9).

$$\begin{aligned}
f(\mathbf{w}) - f(\mathbf{x}^*) &= \frac{1}{2}\mathbf{w}^T H\mathbf{w} + \mathbf{c}^T\mathbf{w} - \frac{1}{2}\mathbf{x}^{*T} H\mathbf{x}^* - \mathbf{c}^T\mathbf{x}^* \\
&= \frac{1}{2}(H\mathbf{w} + \mathbf{c})^T(\mathbf{w} - \mathbf{x}^*) + \frac{1}{2}(H\mathbf{x}^* + \mathbf{c})^T(\mathbf{w} - \mathbf{x}^*) \\
&= \frac{1}{2}(H\mathbf{w} + \mathbf{c} - \alpha\mathbf{b})^T(\mathbf{w} - \mathbf{x}^*) + \frac{1}{2}\alpha\mathbf{b}^T(\mathbf{w} - \mathbf{x}^*) \\
&\qquad - \frac{1}{2}\mu_0\mathbf{x}^{*T}(\mathbf{w} - \mathbf{x}^*) \\
&= -\frac{1}{2}\mu_1\mathbf{w}^T(\mathbf{w} - \mathbf{x}^*) + \frac{1}{2}\alpha\mathbf{b}^T(\mathbf{w} - \mathbf{x}^*) - \frac{1}{2}\mu_0\mathbf{x}^{*T}(\mathbf{w} - \mathbf{x}^*) \\
&= -\frac{1}{2}(\mu_1\mathbf{w} + \mu_0\mathbf{x}^*)^T(\mathbf{w} - \mathbf{x}^*) + \frac{1}{2}\alpha\mathbf{b}^T(\mathbf{w} - \mathbf{x}^*) \\
&= -\frac{1}{2}\mu_0(\mathbf{w} + \mathbf{x}^*)^T(\mathbf{w} - \mathbf{x}^*) - \frac{1}{2}(\mu_1 - \mu_0)\mathbf{w}^T(\mathbf{w} - \mathbf{x}^*) \\
&\qquad + \frac{1}{2}\alpha\mathbf{b}^T(\mathbf{w} - \mathbf{x}^*)
\end{aligned}$$

$$= \frac{1}{2}\mu_0\epsilon - \frac{1}{2}(\mu_1 - \mu_0)\mathbf{w}^T(\mathbf{w} - \mathbf{x}^*) + \frac{1}{2}\alpha\mathbf{b}^T(\mathbf{w} - \mathbf{x}^*)$$
$$= t_1 + t_2 + t_3$$

where

$$t_1 = \frac{1}{2}\mu_0\epsilon,$$
$$t_2 = -\frac{1}{2}(\mu_1 - \mu_0)\mathbf{w}^T(\mathbf{w} - \mathbf{x}^*),$$
$$t_3 = \frac{1}{2}\alpha\mathbf{b}^T(\mathbf{w} - \mathbf{x}^*).$$

Now, each of these terms can be made as small as we want; μ_1 is within $2^{-k}m$ of μ_0, ϵ was chosen above and can be arbitrarily small, and α is also bounded by a multiple of $\mu_1 - \mu_0$.

Appendix: Polynomial root separation

We now provide some background information on polynomials, and then prove Theorem 4.3 via a series of lemmas. For this appendix, the emphasis is on deriving bounds rather than on proposing algorithms. Most of the existence statements, however, can be turned into constructive algorithms. All of the material of this section is explained in greater depth in Zippel [1990].

The set of all polynomials with indeterminate x and coefficients chosen from a ring R is denoted $R[x]$. In many of our examples R will be one of $\mathbb{Z}$, $\mathbb{Q}$, $\mathbb{R}$ or $\mathbb{C}$. A polynomial $f(x)$ is said to be *irreducible* over $R[x]$ if it cannot be written as a product of polynomials of lower degree. The following facts are proved in Jacobson [1974]. First, any polynomial $f(x) \in R[x]$ can be factored as a product of irreducible polynomials for $R = \mathbb{Z}, \mathbb{Q}, \mathbb{R}, \mathbb{C}$. Moreover, this factorization is unique up to constant multiples (where the constant lies in R).

As an application of this theory, suppose $f(x)$ and $g(x)$ are nonzero polynomials over $\mathbb{Q}$ or $\mathbb{R}$ of degree d, e respectively, and suppose there exist polynomials $r(x), s(x)$ such that $r(x)f(x) = s(x)g(x)$, and suppose the degree of $r(x)$ is k such that $k < e$. Then we can conclude that $f(x)$ and $g(x)$ have a common factor. To prove this, let $c_1(x), \ldots, c_i(x)$ be a factorization of $r(x)$ into irreducible polynomials, and let $c_{i+1}(x), \ldots, c_n(x)$ be a factorization of $f(x)$ into irreducibles. Then the sum of degrees of $c_1(x), \ldots, c_i(x)$ must be k and the sum of the degrees of $c_{i+1}(x), \ldots, c_n(x)$ must be d. This is also the (unique) factorization of $s(x)g(x)$, hence $g(x) = \alpha c_{j_1}(x) \cdots c_{j_p}(x)$, where α lies in $\mathbb{Q}$ or $\mathbb{R}$ (depending on which field we started with) and $j_1, \ldots, j_p$ is some subsequence of $1, \ldots, n$. Now, suppose that $\{j_1, \ldots, j_p\} \subset \{1, \ldots, i\}$; this would mean that the degree of g is at most the degree of $c_1(x) \cdots c_i(x) = r(x)$, i.e., the degree is at most k. This contradicts our assumption that $k < e$. Thus, for some index k, $j_k \geq i$. We conclude that $c_{j_k}(x)$ is a common factor to f and g.

Another well known fact about polynomials is that any polynomial $p(x) \in \mathbb{C}[x]$ (and, in particular, any polynomial in $\mathbb{Z}[x]$) can be factored uniquely as $a(x-\alpha_1)\cdots(x-\alpha_d)$ where d is the degree of the polynomial and $\alpha_1, \ldots, \alpha_d$ are the roots of p lying in $\mathbb{C}$.

Next, we prove a lemma taken from Mignotte [1982]. The theorem is due to Cauchy. Mignotte goes on to prove stronger bounds than Theorem 4.3 for root separation, but more advanced techniques are needed.

LEMMA 4.4. *Let $f(x) \in \mathbb{R}[x]$ be the nonzero polynomial*

$$f(x) = c_d x^d + \cdots + c_1 x + c_0.$$

Let $\alpha \in \mathbb{C}$ be a root of f. Then

$$|\alpha| \geq \frac{|c_0|}{|c_0| + \max(|c_1|, \ldots, |c_d|)}.$$

PROOF. If $|\alpha| \geq 1$ then the lemma is trivially true. For the other case, let $m = \max(|c_1|, \ldots, |c_d|)$. By hypothesis α satisfies $c_0 = -c_d\alpha^d - \cdots - c_1\alpha$, so

$$\begin{aligned} |c_0| &\leq m(|\alpha|^d + \cdots + |\alpha|) \\ &\leq m \cdot |\alpha| \cdot (1 + |\alpha| + \cdots + |\alpha|^{d-1}) \\ &\leq \frac{m \cdot |\alpha|}{1 - |\alpha|}. \end{aligned}$$

Noting that $1 - |\alpha| > 0$ by assumption, we can simplify this to get $|\alpha| \geq |c_0|/(|c_0| + m)$, proving the lemma. ∎

The next definition needed for the development of the subject is the resultant.

DEFINITION. *Let $p(x), q(x)$ be polynomials of degree d, e over a field. Suppose that*

$$p(x) = a_d x^d + \cdots + a_0$$

and

$$q(x) = b_e x^e + \cdots + b_0.$$

Then the **resultant** *of $p(x), q(x)$, denoted by $\operatorname{res}_x(p, q)$, is the determinant of* *the following matrix*

$$\underbrace{\begin{bmatrix} a_d & 0 & \cdots & \cdots & 0 \\ a_{d-1} & a_d & \ddots & & \vdots \\ \vdots & a_{d-1} & \ddots & 0 & \vdots \\ a_0 & \vdots & \ddots & a_d & 0 \\ 0 & a_0 & & a_{d-1} & a_d \\ \vdots & 0 & \ddots & \vdots & a_{d-1} \\ \vdots & & \ddots & a_0 & \vdots \\ 0 & \cdots & \cdots & 0 & a_0 \end{bmatrix}}_{e \text{ columns}} \underbrace{\begin{bmatrix} b_e & 0 & \cdots & \cdots & 0 \\ b_{e-1} & b_e & \ddots & & \vdots \\ \vdots & b_{e-1} & \ddots & 0 & \vdots \\ b_0 & \vdots & \ddots & b_e & 0 \\ 0 & b_0 & & b_{e-1} & b_e \\ \vdots & 0 & \ddots & \vdots & b_{e-1} \\ \vdots & & \ddots & b_0 & \vdots \\ 0 & \cdots & \cdots & 0 & b_0 \end{bmatrix}}_{d \text{ columns}}$$

This matrix is known as the **Sylvester matrix** *for p and q.*

We now prove a lemma giving the main property of resultants. This lemma is taken from Jacobson. Although we use resultants only to derive bounds, resultants have been used to construct algorithms. See, for example, Ierardi and Kozen [1990] for a very recent use.

LEMMA 4.5. *If $p(x), q(x)$ are two polynomials over a field, then* $\text{res}_x(p, q) = 0$ *iff p and q have a common factor.*

PROOF. Let A be the Sylvester matrix in the previous definition. We claim this matrix is singular iff p and q have a common factor. Let the common factor be $c(x)$. Write $p(x) = p_1(x)c(x)$ and $q(x) = q_1(x)c(x)$. Suppose the degree of $c(x)$ is k (where $1 \le k \le \min(d, e)$). Write the coefficients of $q_1(x)$ as an $e-k+1$ vector and concatenate $k-1$ zeros. Call this e-vector $\mathbf{x}$. Now write the coefficients of $-p_1(x)$ as a $d-k+1$ vector and concatenate $k-1$ zeros. Call this d-vector $\mathbf{y}$. Now, finally, let $\mathbf{w}$ be the $d+e$ vector resulting from the concatenation of $\mathbf{x}$ and $\mathbf{y}$. We observe that $A\mathbf{w} = \mathbf{0}$. In particular, the ith entry of $A\mathbf{w}$ is the ith coefficient of $x^{k-1}(q_1(x)p(x) - p_1(x)q(x))$ which is the zero polynomial. Thus, the matrix has a nontrivial null-space so it is singular.

Conversely, suppose that matrix A is singular. Let $\mathbf{w}$ be a nontrivial null vector. Then if we split $\mathbf{w}$ into two subvectors, one with e entries and the other with d, these correspond to two polynomials $r(x)$ and $s(x)$ such that $r(x)p(x) + s(x)q(x) = 0$ (identically). Moreover, the degree of r is at most $e-1$ and the degree of s is at most $d-1$, hence $p(x)$ and $q(x)$ have a common multiple (namely $r(x)p(x) = -s(x)q(x)$) of degree at most $d+e-1$. ∎

The following lemma relates differences of roots to resultants. This will allow us to derive the necessary bounds.

LEMMA 4.6. *Let $f(x), g(x)$ be nonzero polynomials in $\mathbb{R}[x]$, and let y be a fixed real number. Then $\text{res}_x(f(x), g(x+y)) = 0$ iff y is the difference between a root of g and a root of f.*

PROOF. This follows directly from the previous lemma. Suppose $y = \alpha - \beta$ such that $g(\alpha) = 0$ and $f(\beta) = 0$. Then $g(x+y)$ and $f(x)$ both have β as a root, therefore their resultant is zero. For the other direction, suppose the resultant is zero. Then $f(x), g(x+y)$ have a common root, say β. This means that $f(\beta) = 0$ and $g(\beta + y) = 0$, i.e., $\beta + y$ is a root of g. Thus, y is a difference between a root of g and a root of f. ∎

For the next theorem, assume that e is at least $3 \log d$.

THEOREM 4.4. *Let $f(x) \in \mathbb{R}[x]$ have degree d and $g(x) \in \mathbb{R}[x]$ degree e. Let $h(y)$ be the function of y defined by $\text{res}_x(f(x), g(x+y))$. Then $h(y)$ is a polynomial function of y of degree at most de. Moreover,*

$$\|h\| \leq 2^{2de} \|f\|^e \|g\|^d.$$

Finally, y_0 is a root of $h(y)$ iff y_0 is a difference between a root of g and a root of f.

PROOF. There are a number of claims to be proved in this theorem. Let us write $f(x) = a_d x^d + \cdots + a_0$ and $g(x) = b_e x^e + \cdots + b_0$. We can express $g(x+y)$ as a polynomial in x with coefficients $c_e(y), \ldots, c_0(y)$. By expanding $g(x+y)$ using the binomial formula we see that $c_i(y)$ is actually a polynomial of degree $e - i$ in y. Moreover, the coefficient of y^j in $c_i(y)$ is given by $b_{i+j}(i+j)!/(i!j!)$. Therefore, the largest magnitude of any coefficient of any c_i is at most $2^e \|g\|$. If we leave y as an indeterminate and write out the matrix corresponding to $\text{res}_x(f(x), g(x+y))$, then the first e columns have coefficients of f and the last d columns have as entries the polynomials $c_0(y), \ldots, c_e(y)$. Since the determinant is a polynomial of its entries, the resultant is a polynomial of y. Moreover, the degree of this polynomial is at most de since the maximum degree with respect to y of any entry is e, and there are d columns containing polynomial entries.

We next compute the largest coefficient of $h(y)$. Since $h(y)$ is a determinant of a $(d+e) \times (d+e)$ matrix, it is a sum of $(d+e)!$ terms; focus on one of these terms. Such a term will be a product of e coefficients of f multiplied by d polynomials chosen from the list $\{c_0(y), \ldots, c_e(y)\}$. An upper bound on the product of e coefficients from f is $\|f\|^e$. Let us consider the coefficient of y^k in this product of d polynomials. The number of terms contributing to this coefficient is the number of ways to partition the integer k into d nonnegative integers; we can put an upper bound of d^d on the number of terms. Each term is the product of at most k coefficients of a particular c_i; each of these coefficients has $\|g\| 2^e$ as an upper bound. Thus, the term is at most $\|g\|^d 2^{de}$; the number of terms is d^d contributing to y^k, and the number of terms contributing to the determinant is $(d+e)!$. This gives an upper bound of $(d+e)!d^d 2^{de}\|f\|^e\|g\|^d$. We can estimate the first three factors as 2^{2de} provided that e is larger than $3\log d$. This gives the bound stated in the theorem.

Now, finally, we turn to the last claim made in the theorem. That claim follows from the previous lemma, because $h(y_0) = 0$ if and only if $\text{res}_x(f(x), g(x+y_0)) = 0$, i.e., if and only if y_0 is the difference between a root of g and a root of f. ∎

We are now in a position to prove Theorem 4.3 stated in the body of this section.

PROOF OF THEOREM 4.3. Apply the previous theorem to get a polynomial $h(y)$ of degree de that has $\beta - \alpha$ as root and has coefficients at most $2^{2de}\|f\|^e\|g\|^d$. Divide this polynomial by powers of y until the constant term becomes nonzero; call this new polynomial h_1. Polynomial h_1 will still have $\beta - \alpha$ as a root and will have the same bound on the coefficient size. Note that the coefficients of h_1 are integers because they are polynomials in the coefficients of f and g.

Now we apply Lemma 4.4 to conclude that all roots of polynomial $h_1(y)$ have magnitude at least $a/(a+b)$, where a is the magnitude of the constant coefficient of h_1 and b is the maximum magnitude of the other coefficients. We know that $a \geq 1$ since h_1 has integer coefficients, and $a, b \leq 2^{2de}\|f\|^e\|g\|^d$. This is a lower bound on $|\beta - \alpha|$, proving the theorem. ∎

4.4. Algorithms for indefinite quadratic problems

In light of the results in this chapter, we cannot hope to get an efficient algorithm for the general case of nonconvex quadratic programming. In this section we propose a straightforward enumerative algorithm for the problem. This algorithm has exponential running time.

To derive an enumerative algorithm for the nonconvex problem

$$\begin{aligned} &\text{minimize} && \tfrac{1}{2}\mathbf{x}^T H\mathbf{x} + \mathbf{c}^T\mathbf{x} \\ &\text{subject to} && A\mathbf{x} \geq \mathbf{b} \end{aligned} \tag{4.12}$$

we return to the proof of Theorem 4.1. Recall that there were two cases for this proof: Either the minimum exists or it fails to exist. If the minimum exists, then it is equal to the solution of a system of linear equations, as proved in the last chapter. These equations have the form

$$\begin{pmatrix} H & -M^T \\ M & 0 \end{pmatrix} \begin{pmatrix} \mathbf{x} \\ \boldsymbol{\mu} \end{pmatrix} = \begin{pmatrix} -\mathbf{c} \\ \mathbf{b}' \end{pmatrix} \tag{4.13}$$

where $(M, \mathbf{b}')$ is composed of a subset of rows of the system $A\mathbf{x} = \mathbf{b}$ combined with a subset of rows of the system $\mathbf{x} = \mathbf{0}$. Therefore, we can find a global minimum as follows. Set $(M, \mathbf{b}')$ sequentially to all possible subsets of the equations:

$$\begin{pmatrix} A \\ I \end{pmatrix} \mathbf{x} = \begin{pmatrix} \mathbf{b} \\ \mathbf{0} \end{pmatrix}.$$

For each choice of $(M, \mathbf{b}')$ solve (4.13) (provided it is nonsingular). For the solution vector $\mathbf{x}$, check whether $A\mathbf{x} \geq \mathbf{b}$ is satisfied, and, if it is, compute the objective function value $\frac{1}{2}\mathbf{x}^T H\mathbf{x} + \mathbf{c}^T\mathbf{x}$. Find the minimum such objective function value.

The running time for this algorithm is seen to be $O(2^{m+n}(m+n)^3)$ arithmetic operations because there are 2^{m+n} choices for M and each choice requires the solution of a system of equations. Clearly there are ways to make this algorithm more efficient, but any approach like this would seem to require exponential dependence on m or n.

Next, we must address the case that the minimum does not exist. The minimum may not exist because $A\mathbf{x} \geq \mathbf{b}$ is infeasible; this can be checked in polynomial time using the methods of Chapter 3. The other case is that the objective function is unbounded from below. This can also be tested via an exponential-time algorithm. In particular, we enumerate all possible "signatures" as in Section 4.1, and then solve the appropriate systems of equations.

Another approach for the general problem is the class of partitioning algorithms. For these algorithms the feasible region is subdivided, and a convex approximation to the objective function is formulated for each subregion. A branch-and-bound approach (see Papadimitriou and Steiglitz [1982] for the definition of *branch and bound*) can be formulated to determine which of the subregions should be further subdivided. This class of methods is fully described in Pardalos and Rosen [1987].

Exercises

4.1 Consider the decision problem of determining whether a region specified by quadratic constraints is nonempty (see Exercise 2.5). It is not known whether this problem lies in **NP**. Give an example of a domain specified by quadratic constraints such that every rational point in the domain requires an exponential number of digits to write.

4.2 In Section 4.1 we established that a quadratic function was either unbounded or attained its minimum on a nonempty region of the form $A\mathbf{x} \geq \mathbf{b}$. Consider the case of minimizing a linear function on a region given by quadratic constraints as in the previous problem. Show that in this case it is possible for the objective function to be bounded below but for the optimum not to be attained. [Note: There is an example with two variables.]

4.3 Show that the following problem is **NP**-hard. Given H, minimize $\frac{1}{2}\mathbf{x}^T H\mathbf{x}$ subject to $-1 \leq x_i \leq 1$ for $i = 1, \ldots, n$.

4.4 Give explicit examples in $\mathbb{R}^2$ of minimizing $f(\mathbf{x}) = \frac{1}{2}\mathbf{x}^T H\mathbf{x} + \mathbf{c}^T\mathbf{x}$ on a disk such that (a) $\mu_0 < 0$ (b) $\mu_0 = \mu^*$ (c) $\mu^* > \mu_0 > 0$. [Hint: It is easiest if you work with a 2×2 diagonal H.]

4.5 In the algorithm for minimizing a nonconvex quadratic over a sphere, it is necessary to repeatedly solve systems of the form $H + \mu I$ for varying values of μ. Show that these steps can be sped up by factoring H at the beginning of the algorithm into the form QTQ^T, where Q is orthogonal and T is symmetric tridiagonal. [Note: This technique is used for the symmetric QR algorithm for eigenvalue computations; it is necessary to solve systems of this form to recover eigenvectors. See Golub and Van Loan [1989] for more information.]

In the next three exercises, we present some of the ideas necessary to prove that the decision version of sphere-constrained quadratic programming lies in **P**. The decision problem is: Given $(H, \mathbf{c}, \zeta)$, is there an $\mathbf{x} \in \mathbb{R}^n$ such that $\mathbf{x}^T\mathbf{x} \leq 1$ and $\frac{1}{2}\mathbf{x}^T H\mathbf{x} + \mathbf{c}^T\mathbf{x} \leq \zeta$? These exercises do not lead to a proof because many lengthy details have been omitted. For the full proof, see Vavasis and Zippel [1990].

4.6 Let α be a root of a polynomial $p(x)$, where $p(x)$ has degree at most d and integer coefficients of magnitude at most s. Let $\theta = q(\alpha)/r(\alpha)$, where $q(x), r(x)$ are polynomials of degree at most e with integer coefficients of magnitude at most t. Assume that p, q, r have no common roots. Show that θ is the root of a nonzero polynomial with integer coefficients. Derive an upper bound on the degree of this polynomial and on the magnitude of its coefficients. [Hint: Consider the polynomial $q(x) - yr(x)$. For what values of y does this polynomial have a common root with $p(x)$?]

4.7 Suppose we have an instance of (4.4) with $\mu^* > \mu_0$. Let θ be the optimal value of the objective function. Show that θ can be expressed as $r(\mu^*)/s(\mu^*)$, where $r(x), s(x)$ are polynomials with integer coefficients. What is an upper bound on the degree of these polynomials and on the magnitude of their coefficients?

4.8 Let θ be the optimal value of the objective function to (4.4), and suppose θ is a root to an integer polynomial of degree d with coefficients at most s in magnitude. Let ζ be a given rational number with numerator and denominator of magnitude at most t. Suppose an approximation θ_1 of θ is computed, such that $|\theta_1 - \theta| \leq 2^{-k}$. How large must k be in order to determine from θ_1 whether $\theta = \zeta$, $\theta < \zeta$, or $\theta > \zeta$?

5

LOCAL OPTIMIZATION

In the last chapter we saw that many versions of nonconvex quadratic programming are **NP**-hard. Indeed, practitioners have realized for years that global minimization of nonconvex problems seems computationally intractable. Accordingly, most well-known optimization packages (e.g., MINOS by Murtagh and Saunders [1987]) try to produce only local optima. In this chapter we investigate the problem of local optimization from a complexity-theoretic point of view. In the first section, we show that the general problem of local optimality for nonconvex QP is an **NP**-hard problem. This result seems surprising since it has been tacitly assumed that local optima are easy to find.

The second section characterizes local minima for QP. In the third, fourth and fifth sections we address the problem of local minima for quadratic knapsack problems. We show that this problem can be solved in polynomial time.

The practitioner of optimization ought to read this chapter with the question, "Is it reasonable to expect that the termination point of an optimization algorithm be a local minimum?" The answer to this question is not fully understood, but in this chapter we provide some positive and negative results.

5.1. General quadratic local minimization

We define QPLOC to be the following decision problem. Given an instance $(H, \mathbf{c}, A, \mathbf{b})$ of quadratic programming, and given a point $\mathbf{x}^* \in \mathbb{R}^n$, is $\mathbf{x}^*$ a local minimum of the problem? In other words, (1) does $\mathbf{x}^*$ satisfy $A\mathbf{x}^* \geq \mathbf{b}$, and (2) if so, does there exist an $\epsilon > 0$ such that $\frac{1}{2}\mathbf{x}^T H\mathbf{x} + \mathbf{c}^T\mathbf{x} \geq \frac{1}{2}\mathbf{x}^{*T} H\mathbf{x}^* + \mathbf{c}^T\mathbf{x}^*$ for all $\mathbf{x}$ satisfying $\|\mathbf{x} - \mathbf{x}^*\|_2 \leq \epsilon$ and $A\mathbf{x} \geq \mathbf{b}$?

The main result of this section is the following theorem. This theorem is due to Murty and Kabadi [1987], although our proof is a simplification because we can take advantage of the Motzkin-Strauss results proved in the last chapter.

THEOREM 5.1. *Problem QPLOC is* **NP**-*hard.*

PROOF. Let G be an undirected graph. We show that the problem of determining whether G has a clique of size k can be reduced to an instance of QPLOC. Accordingly, assume k is given. Define $f(\mathbf{x})$ to be the quadratic objective function given by (4.2) based on graph G. Notice that this objective function may be written as $\mathbf{x}^T H_0 \mathbf{x}$ (no constant or linear term). Define $g(\mathbf{x})$ to be $f(\mathbf{x}) - c$ where

$$c = \frac{1}{2k-1} - \frac{1}{2}.$$

Notice that if the maximum clique size is k, then $g(\mathbf{x})$ will attain a negative value on Δ^{n-1}; otherwise it will always be positive on Δ^{n-1}.

Next, notice that if $\mathbf{x} \in \Delta^{n-1}$ then c is the same as $c(x_1 + \cdots + x_n)^2$. Therefore, we can introduce objective function $h(\mathbf{x})$ defined to be

$$h(\mathbf{x}) = f(\mathbf{x}) - c(x_1 + \cdots + x_n)^2;$$

again, this objective function will attain a negative value on Δ^{n-1} iff G has a clique of size at least k.

Notice that $h(\mathbf{x})$ takes the form $\mathbf{x}^T H \mathbf{x}$ (no constant or linear term). Therefore, we have produced a matrix H such that $\mathbf{x}^T H \mathbf{x}$ is negative on Δ^{n-1} if and only if G has a clique of size at least k.

Let Q be the orthant $\{\mathbf{x} \in \mathbb{R}^n : \mathbf{x} \geq \mathbf{0}\}$. Let a be any positive real number. Consider the problem of minimizing $\mathbf{x}^T H \mathbf{x}$ subject to the constraints $\mathbf{x} \in Q$ and $\mathbf{e}^T \mathbf{x} = a$, where $\mathbf{e}$ is the vector of all 1's. For $a = 1$ these constraints specify Δ^{n-1}. Notice, however, that objective function values scale exactly proportionally to a^2. Therefore, for any fixed choice of a, we can claim that $\mathbf{x}^T H \mathbf{x}$ is negative subject to the preceding constraints if and only if G has a clique of size at least k.

Now, finally, consider the problem of minimizing $\mathbf{x}^T H \mathbf{x}$ subject only to $\mathbf{x} \in Q$. We claim that $\mathbf{0}$ is a local (and, in fact, global) minimum of $h(\mathbf{x})$ on this region if and only if G does not have clique of size at least k. This result would prove the theorem. If G does not have a clique of size k, then the preceding argument shows that $h(\mathbf{x})$ is nonnegative at every point of Q. On the other hand, if G has a clique of size k, then there exists a point $\mathbf{x} \in Q$ such that $h(\mathbf{x}) < 0$, and $\mathbf{e}^T \mathbf{x}$ may be arbitrarily small (but positive). This means that $\mathbf{0}$ cannot be a local minimum. ∎

5.2. Characterizing local minima for quadratic programs

In this section we provide characterizations of local minima of quadratic programming problems. These characterizations are useful for the upcoming sections of this chapter.

We start by giving *second order* conditions for quadratic programming problem:

$$\begin{aligned} \text{minimize} \quad & f(\mathbf{x}) = \tfrac{1}{2}\mathbf{x}^T H\mathbf{x} + \mathbf{c}^T\mathbf{x} \\ \text{subject to} \quad & A\mathbf{x} \geq \mathbf{b}. \end{aligned} \tag{5.1}$$

Let $\mathbf{x}^*$ be a feasible point in $\mathbb{R}^n$. Let $\mathbf{g}$ denote $H\mathbf{x}^* + \mathbf{c}$, the gradient of f at $\mathbf{x}^*$. Let $(A_U, \mathbf{b}_U)$ denote the subset of constraints active at $\mathbf{x}^*$, so that $A_U\mathbf{x}^* = \mathbf{b}_U$. We claim that the following two conditions are necessary and sufficient for local minimality of (5.1):

1. For all $\mathbf{d}$ such that $A_U\mathbf{d} \geq \mathbf{0}$, $\mathbf{g}^T\mathbf{d} \geq 0$.

2. For all $\mathbf{d}$ such that $A_U\mathbf{d} \geq \mathbf{0}$ and $\mathbf{g}^T\mathbf{d} = 0$, $\mathbf{d}^T H\mathbf{d} \geq 0$.

Observe that $\mathbf{d}$ is a feasible direction at $\mathbf{x}^*$ if and only if $A_U\mathbf{d} \geq \mathbf{0}$. Note that Condition 1 combined with feasibility is equivalent to the KKT conditions. Condition 2 is the additional *second order condition.*

Recall that a *feasible direction* for an optimization problem at a feasible point $\mathbf{x}^*$ is a vector $\mathbf{d}$ such that there is an $\epsilon > 0$ such that $\mathbf{x}^* + t\mathbf{d}$ is feasible for all $t \in [0, \epsilon]$. A *descent direction* is a vector $\mathbf{d}$ such that there is an $\epsilon > 0$ such that $f(\mathbf{x}^* + t\mathbf{d}) < f(\mathbf{x}^*)$ for all $t \in (0, \epsilon)$. Finally, a vector $\mathbf{d}$ is called a *feasible descent direction* if it is both a feasible direction and a descent direction. Clearly, the presence of a feasible descent direction means that $\mathbf{x}^*$ is not a local minimum.

Before proving that Conditions 1 and 2 are equivalent to local minimality, we first prove the following simpler auxiliary theorem that gives an alternative characterization of the conditions.

THEOREM 5.2. *Conditions* 1 *and* 2 *are equivalent to the condition that there are no feasible descent directions at* $\mathbf{x}^*$.

PROOF. First, suppose Conditions 1 and 2 hold. Then we claim that there are no feasible descent directions. Let $\mathbf{d}$ be a feasible direction at $\mathbf{x}^*$, so that $A_U\mathbf{d} \geq \mathbf{0}$. Then

$$f(\mathbf{x}^* + t\mathbf{d}) - f(\mathbf{x}^*) = t\mathbf{g}^T\mathbf{d} + t^2\mathbf{d}^T H\mathbf{d}/2. \tag{5.2}$$

Condition 1 implies that $\mathbf{g}^T\mathbf{d} \geq 0$. If $\mathbf{g}^T\mathbf{d} > 0$, then the first term of the right-hand side of (5.2) is positive and dominates the second term for t small enough, so $\mathbf{d}$ cannot be a descent direction. If $\mathbf{g}^T\mathbf{d} = 0$ then the second term is nonnegative by Condition 2, hence $\mathbf{d}$ once again is not a descent direction.

Conversely, suppose Condition 1 or Condition 2 fail. Then it is easily seen that the $\mathbf{d}$ that violates either condition is a feasible descent direction. ∎

We now prove the main theorem for this section. This theorem is due to Contesse [1980], but our proof is different.

THEOREM 5.3. *Conditions* 1 *and* 2 *are necessary and sufficient for a feasible point* $\mathbf{x}^*$ *to be a local minimum of* (5.1).

PROOF. The necessity of these conditions is clear. Any $\mathbf{d}$ that violated either condition is a feasible descent direction at $\mathbf{x}^*$.

For the other direction, consider the parameterized optimization problem:

$$r(\lambda) = \min\{\mathbf{d}^T H\mathbf{d} : A_U\mathbf{d} \geq \mathbf{0}; \|\mathbf{d}\|_\infty \leq 1; \mathbf{g}^T\mathbf{d} \leq \lambda\}. \tag{5.3}$$

Notice that the optimization problem on the right-hand side of (5.3) is feasible for all $\lambda \geq 0$ since $\mathbf{d} = \mathbf{0}$ is feasible. Therefore, the function $r(\lambda)$ is well-defined for all $\lambda \geq 0$. It can be proved (see Exercise 1.2) that $r(\lambda)$ is continuous. Notice also that $r(\lambda) \leq 0$ for all $\lambda \geq 0$, again, because the vector $\mathbf{0}$ is feasible for this problem. On the other hand, $r(0) \geq 0$ because of Conditions 1 and 2. In particular, any $\mathbf{d}$ satisfying $A_U\mathbf{d} \geq \mathbf{0}, \mathbf{g}^T\mathbf{d} \leq 0$ must actually satisfy $\mathbf{g}^T\mathbf{d} = 0$ by Condition 1, and therefore satisfies $\mathbf{d}^T H\mathbf{d} \geq 0$ by Condition 2. Therefore, $r(0) = 0$.

The goal for the next two paragraphs is to show that there exists a number $\lambda_1 > 0$ such that r is a quadratic function on the interval $[0, \lambda_1]$. Fix a choice of $\lambda \geq 0$. Let $\mathbf{d}^*(\lambda)$ be a global minimum of the quadratic program in (5.3). If there is more than one global minimum, then we follow the rules in Chapter 3. Specifically, choose the global minimum with the highest number of active constraints. Break further ties by selecting the global minimum with the highest number of coordinates equal to zero. If there are still ties, they are broken arbitrarily. In this case, $\mathbf{d}^*(\lambda)$ will be the solution to a nonsingular system of linear equations whose coefficients come from the coefficients of (5.3) and whose right-hand side depends linearly on λ.

There are a finite number of possible combinatorial choices for producing that matrix that has $\mathbf{d}^*(\lambda)$ as its solution. Specifically, there are a finite number of ways to choose which constraints are active and which coordinates are equal to zero. Consider all such combinatorial choices that yield linear systems whose solutions are feasible in a small enough closed set $[0, \lambda_0]$. Among these combinatorial choices, there will be one whose objective function is minimum for some small enough interval $[0, \lambda_1]$, $\lambda_1 > 0$. This choice will define the optimal $\mathbf{d}^*(\lambda)$ for λ near zero. Thus, $\mathbf{d}^*(\lambda)$ can be taken to be a linear function of λ in interval $[0, \lambda_1]$. This means that $r(\lambda)$ is quadratic in this interval.

As stated above, $r(0) = 0$, hence this quadratic has the form

$$r(\lambda) = a\lambda^2 + b\lambda$$

for $\lambda \in [0, \lambda_1]$. Because $r(\lambda)$ has no constant term, there exists a number $t > 0$ such that

$$r(\lambda) \geq -t\lambda$$

for all $\lambda \in [0, \lambda_1]$.

Define

$$\epsilon = \min(2\lambda_1/\|H\|_\infty, 2/t).$$

Consider a feasible direction $\mathbf{d}$ such that $\|\mathbf{d}\|_\infty \le \epsilon$. We take two cases; either $\mathbf{g}^T\mathbf{d} \ge \|\mathbf{d}\|_\infty \cdot \lambda_1$, or $\mathbf{g}^T\mathbf{d} < \|\mathbf{d}\|_\infty \cdot \lambda_1$. In the first case,

$$\begin{aligned} f(\mathbf{x}^* + \mathbf{d}) &= f(\mathbf{x}^*) + \mathbf{g}^T\mathbf{d} + \mathbf{d}^T H\mathbf{d}/2 \\ &\ge f(\mathbf{x}^*) + \|\mathbf{d}\|_\infty \cdot \lambda_1 - \epsilon \cdot \|\mathbf{d}\|_\infty \cdot \|H\|_\infty /2 \\ &\ge f(\mathbf{x}^*) + \|\mathbf{d}\|_\infty \cdot (\lambda_1 - \epsilon \cdot \|H\|_\infty /2) \\ &\ge f(\mathbf{x}^*). \end{aligned}$$

In the second case, let $\lambda = \mathbf{g}^T\mathbf{d}/\|\mathbf{d}\|_\infty$. Then $\lambda < \lambda_1$. In this case,

$$\mathbf{d}^T H\mathbf{d} \ge \|\mathbf{d}\|_\infty^2 \, r(\lambda)$$

since $\mathbf{d}/\|\mathbf{d}\|_\infty$ is feasible for the optimization problem defining $r(\lambda)$. Thus, $\mathbf{d}^T H\mathbf{d} \ge -t\lambda \|\mathbf{d}\|_\infty^2$. We have the following computation:

$$\begin{aligned} f(\mathbf{x}^* + \mathbf{d}) &= f(\mathbf{x}^*) + \mathbf{g}^T\mathbf{d} + \mathbf{d}^T H\mathbf{d}/2 \\ &\ge f(\mathbf{x}^*) + \lambda \cdot \|\mathbf{d}\|_\infty - t\lambda\epsilon \cdot \|\mathbf{d}\|_\infty /2 \\ &\ge f(\mathbf{x}^*) + \lambda \cdot \|\mathbf{d}\|_\infty \cdot (1 - t\epsilon/2) \\ &\ge f(\mathbf{x}^*) \end{aligned}$$

by choice of ϵ. ∎

The preceding two theorems immediately give us the following corollary:

COROLLARY 5.1. *A point* $\mathbf{x}^*$ *is a local minimum of* (5.1) *if and only if there are no feasible descent directions at* $\mathbf{x}^*$.

We remark that this corollary does not hold even for simple generalizations of (5.1). For example, consider the problem of minimizing $y - x^2$ subject to $x^2 + y^2 \le 1$. It is an exercise (Exercise 5.2) that the point $(0, -1)$ is not a local minimum, even though there are no feasible descent directions at this point. Fig. 5.1 shows the feasible region and the level curve passing through $(0, -1)$.

5.3. Local minima for concave knapsack problems

In light of the theorem of Section 5.1, one might ask under what circumstances it is tractable to compute a local minimum of a quadratic program. Clearly if the problem is convex, then a local minimum can be found in polynomial time since a global minimum is a local minimum. Accordingly, the interesting question is whether a polynomial-time algorithm can find

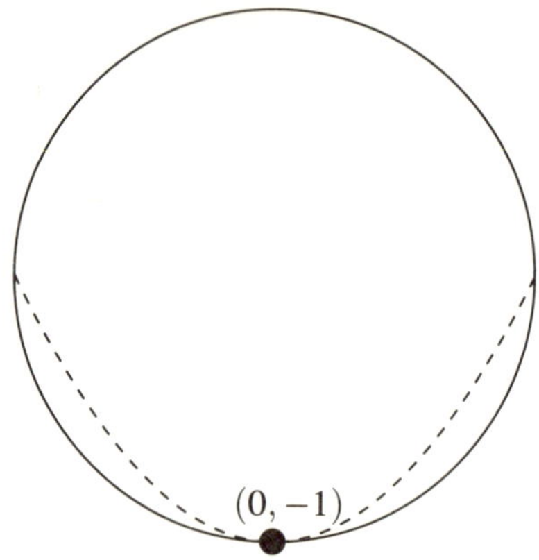

FIG. 5.1. Point $(0, -1)$ has no feasible descent directions.

local minima for a class of quadratic programming problems whose global minimization is **NP**-hard.

This problem is the topic of this and the next two sections. In particular, we show that a local minimum for the quadratic knapsack problem (QKP) introduced in Chapter 3 can be found in polynomial time. The results of this section and the next section are due to Moré and Vavasis [1991].

Recall that QKP takes the following form.

$$\begin{aligned} \text{minimize} \quad & \tfrac{1}{2}\mathbf{x}^T D\mathbf{x} + \mathbf{c}^T\mathbf{x} \\ \text{subject to} \quad & \mathbf{a}^T\mathbf{x} = \gamma, \\ & l_i \le x_i \le u_i \quad \text{for } i = 1, \ldots, n. \end{aligned} \tag{5.4}$$

Here, D is a diagonal matrix. As in Chapter 3, without loss of generality we can assume the equation constraint has the form $x_1 + \cdots + x_n = \gamma$. In Chapter 4 we argued that this problem is **NP**-hard in the case that D is negative definite. In this section and the next we will treat the local minimization problem for the negative definite case. In the section afterwards we will turn to the indefinite case. As in Chapter 3, we rewrite (5.4) as

$$\begin{aligned} \text{minimize} \quad & q_1(x_1) + \cdots + q_n(x_n) \\ \text{subject to} \quad & x_1 + \cdots + x_n = \gamma, \\ & l_i \le x_i \le u_i \quad \text{for } i = 1, \ldots, n \end{aligned} \tag{5.5}$$

where each $q_i(x)$ is a strictly concave quadratic function of one variable. We say that a function $f : \mathbb{R}^n \to \mathbb{R}$ is *concave* if $-f$ is a convex function. As in Chapter 3, we write $q_i(x_i)$ as $\frac{1}{2}d_i x_i^2 + c_i x_i$, but now the assumption is $d_i < 0$. We also assume that $l_i < u_i$. If $l_i = u_i$ then x_i is uniquely determined and may be dropped from the problem. If $l_i > u_i$ then the problem is infeasible.

In order to motivate our algorithm, we establish some lemmas to characterize the local minimum. We first establish the following simple lemma about the negative definite case.

LEMMA 5.1. *Let* $\mathbf{x}^*$ *be a local minimum to a negative definite instance of* (5.5). *Then* $l_i < x_i < u_i$ *can hold for at most one variable* x_i^*.

PROOF. This lemma follows from a general principal concerning strictly concave functions, but we give a direct proof. Suppose the lemma is false, e.g., suppose that $x_1^* \in (l_1, u_1)$ and $x_2^* \in (l_2, u_2)$. Then for $\epsilon > 0$ chosen small enough, the points

$$\mathbf{x}' = (x_1^* - \epsilon, x_2^* + \epsilon, x_3^*, \ldots, x_n^*)$$

and

$$\mathbf{x}'' = (x_1^* + \epsilon, x_2^* - \epsilon, x_3^*, \ldots, x_n^*)$$

are both feasible. If $f(\mathbf{x})$ denotes the objective function, then

$$f(\mathbf{x}') = f(\mathbf{x}^*) - d_1 x_1^* \epsilon + \frac{1}{2} d_1 \epsilon^2 + d_2 x_2^* \epsilon + \frac{1}{2} d_2 \epsilon^2 - c_1 \epsilon + c_2 \epsilon$$

and

$$f(\mathbf{x}'') = f(\mathbf{x}^*) + d_1 x_1^* \epsilon + \frac{1}{2} d_1 \epsilon^2 - d_2 x_2^* \epsilon + \frac{1}{2} d_2 \epsilon^2 + c_1 \epsilon - c_2 \epsilon.$$

Therefore, the average of $f(\mathbf{x}')$ and $f(\mathbf{x}'')$ is $f(\mathbf{x}^*) + \frac{1}{2}(d_1 + d_2)\epsilon^2$. The term $(d_1 + d_2)\epsilon^2$ is negative since D is negative definite. In particular, this means that one of $f(\mathbf{x}')$ and $f(\mathbf{x}'')$ is less than $f(\mathbf{x}^*)$. Since this is true for all $\epsilon > 0$ small enough, local optimality is violated. ∎

The preceding lemma tells us that at a local minimum, all but one of the variables is at its bound. Thus, the selection of the local minimum reduces to the combinatorial problem of determining which variables should be set to upper or lower bounds.

Next, we characterize a local minimum of (5.5).

LEMMA 5.2. *Let* $\mathbf{x}^*$ *be a feasible point such that at most one variable is not at a lower or upper bound. Then the following conditions are necessary and sufficient for local minimality. There exists a number* λ *such that*

1. *If* $l_i < x_i^* < u_i$ *then* $q_i'(x_i^*) = \lambda$.
2. *If* $x_i^* = l_i$ *then* $q_i'(x_i^*) > \lambda$.
3. *If* $x_i^* = u_i$ *then* $q_i'(x_i^*) < \lambda$.

REMARKS.

1. These three inequalities represent a strengthening of the KKT conditions; see (3.3) for the KKT conditions. Since (5.5) is not a convex problem, the KKT conditions are not sufficient for local minimality, but these stronger conditions are indeed sufficient.

2. Necessity of these conditions is not proved here but is left to the reader as Exercise 5.3. Only sufficiency is used in the upcoming material.

PROOF. Since (5.5) is a quadratic program, we can apply Theorem 5.3 to show that Conditions 1–3 imply that there are no feasible descent directions. Let $\mathbf{d}$ be a nonzero feasible direction at a feasible point $\mathbf{x}^*$ satisfying the conditions of the lemma. Let $\mathbf{g} = \nabla f(\mathbf{x}^*)$. We claim that $\mathbf{g}^T\mathbf{d} > 0$; this would give us local minimality. Since $\mathbf{d}$ is feasible, $\mathbf{e}^T\mathbf{d} = 0$, where $\mathbf{e}$ denotes the vector of all 1's. Therefore,

$$\mathbf{g}^T\mathbf{d} = (\mathbf{g} - \lambda\mathbf{e})^T\mathbf{d}.$$

Choose an index i, and look at the ith contribution to the inner product $(\mathbf{g} - \lambda\mathbf{e})^T\mathbf{d}$. This contribution is equal to $(q_i'(x_i^*) - \lambda)d_i$. Since $d_i \geq 0$ for all i such that $x_i^* = l_i$, and $d_i \leq 0$ for all i such that $x_i^* = u_i$, we see that the conditions specified in the lemma imply that this contribution is nonnegative.

In fact, we claim that this inner product is positive if $\mathbf{d} \neq \mathbf{0}$. Since $\mathbf{e}^T\mathbf{d} = 0$, there must be at least two coordinates of $\mathbf{d}$ with opposite signs that are nonzero; say $d_1 > 0$ and $d_2 < 0$. This means that either $x_1^* = l_1$ or $x_2^* = u_2$. (One of the two variables could be strictly between l_i and u_i, in which case d_i is not restricted in sign, but at most one could be strictly between.) Suppose, e.g., that $x_1^* = l_1$. Checking the conditions indicates that the first contribution to the inner product $(\mathbf{g} - \lambda\mathbf{e})^T\mathbf{d}$ is therefore positive.

Thus, if $\mathbf{d}$ is a feasible direction at $\mathbf{x}^*$ satisfying the conditions of the lemma, it cannot be a descent direction.

Therefore, we conclude that the conditions of the lemma imply local minimality. ∎

Suppose we were trying to formulate an algorithm that would set $x_i^* = l_i$ for some sequence of indices i in $\{1, \ldots, n\}$ in order to make a local minimum. The previous lemma suggests that, among all choices of i, we should choose the one such that $q_i'(l_i)$ is largest; heuristically, it seems that this choice would give the most freedom when it comes time to pick λ to satisfy the lemma.

For an arbitrary i, it is not always possible to set $x_i^* = l_i$ and maintain feasibility with respect to the other constraints. Suppose we have a set F of indices such that we have decided upon values for x_i^* for all $i \in F$, and we have not decided how to set variables x_i^* for $i \in \{1, \ldots, n\} - F$. For each

$i \notin F$, define a quantity

$$r_i = \gamma - \sum_{j \in F} x_j^* - \sum_{j \notin F} u_j + u_i. \tag{5.6}$$

We claim that r_i is the smallest possible setting of x_i^* consistent with the choices made so far and consistent with the equation constraint. We will not delve into a proof of this claim, since it is implicit in the proofs of some of the upcoming lemmas. Note that $u_i - r_i$ is a quantity that does not depend on i.

Accordingly, we propose the following algorithm to compute a local minimum. After stating the algorithm, we shall prove a series of lemmas establishing that indeed it computes a local minimum. In the next section, we investigate efficient implementations.

ALGORITHM LOCAL-MIN1
- Let $F^{(1)} := \emptyset$.
- for $k := 1, \ldots, n$
  - For each $i \notin F^{(k)}$, compute r_i according to (5.6).
  - Find $i \notin F^{(k)}$ such that $q_i'(\max(l_i, r_i))$ is maximum.
  - Let $x_i^* := \max(l_i, r_i)$.
  - Let $F^{(k+1)} := F^{(k)} \cup \{i\}$.
- end for.

On the step of finding the maximum, ties are broken arbitrarily. An obvious property of this algorithm is that F increases by one index each iteration, and that after the final iteration F includes all indices $\{1, \ldots, n\}$. In addition, each x_i^* is set to a value by this algorithm.

We now establish that the vector $\mathbf{x}^*$ computed by this algorithm is, first, feasible, and second, a local minimum.

Focus on a particular iteration k. The superscript '$-$' indicates values of the variables at the previous iteration of the algorithm, and the superscript '$+$' indicates values at the next iteration.

LEMMA 5.3. *Suppose the initial problem is feasible. Assume $i \notin F$. Then $r_i \geq r_i^-$ and $r_i \leq u_i$.*

PROOF. We prove this by induction on the number of steps. For the base case, we examine the first step. At the first step

$$r_i = \gamma - \sum_{j \neq i} u_j.$$

Hence, if $r_i > u_i$ at this step, then $u_1 + \cdots + u_n < \gamma$ which implies infeasibility. The other claim does not apply to the first step.

Next, we turn to the induction case; assume we are at a step $k > 1$. We denote the values of r_i at the previous iteration by r_i^-; the induction assumption is that the lemma was valid at the previous step. Let ξ be the unique element in $F - F^-$. Suppose $i \notin F$. Notice that $r_i - r_i^- = u_\xi - x_\xi^*$. By definition $x_\xi^* = \max(l_\xi, r_\xi^-)$. We know that $l_\xi < u_\xi$ and, by induction, $r_\xi^- \le u_\xi$. Therefore, $r_i - r_i^- \ge 0$, proving the first statement of the lemma. Now we prove the second statement. We have the following sequence of calculations.

$$\begin{aligned} u_i - r_i &= \sum_{j \in F} x_j^* + \sum_{j \notin F} u_j - \gamma \\ &= \sum_{j \in F^-} x_j^* + x_\xi^* - u_\xi + \sum_{j \notin F^-} u_j - \gamma \\ &\ge \sum_{j \in F^-} x_j^* + r_\xi^- - u_\xi + \sum_{j \notin F^-} u_j - \gamma. \end{aligned} \tag{5.7}$$

Now, if we substitute formula (5.6) for r_ξ^- we notice that the entire right hand side cancels, and we are left with 0. ∎

We now try to give a more accurate description of the variables x_i^* returned by the algorithm.

LEMMA 5.4. *Let m be the first iteration of Algorithm LOCAL-MIN1 such that $x_i^* = r_i$ and $r_i > l_i$. Then on all iterations preceding m, x_i^* is set to l_i by the algorithm. On all iterations following m, x_i^* is set to u_i.*

PROOF. Let m be as in the lemma. The statement that x_i^* is set to l_i during all previous iterations follows trivially from the definition of m.

To prove the other case, examine iteration $m + 1$ of the algorithm. Consider (5.7) used in the preceding proof. We notice that $x_\xi^* = r_\xi^-$ by choice of m, and therefore the inequalities in (5.7) are satisfied as equations. Since the final right-hand side is 0, we conclude that $u_i = r_i$ for all $i \notin F$. But this equation must necessarily hold at all successive iterations, since no r_i can decrease nor can it increase beyond u_i according to the last lemma. ∎

LEMMA 5.5. *Assume that the original problem (5.5) is feasible. Then the point returned by Algorithm LOCAL-MIN1 is feasible.*

PROOF. By Lemma 5.3, we know that $x_i^* \le u_i$ for all i. We also know directly from the algorithm that $x_i^* \ge l_i$ for all i. The only remaining constraint is the equation. Let ξ be the index added to F in the last iteration of the algorithm. Then, by definition, at this iteration

$$r_\xi = \gamma - \sum_{j \ne \xi} x_j^*.$$

Accordingly, if x_ξ^* were set to r_ξ, then the equation constraint would hold.

The other case is that $x_\xi^* \ne r_\xi$, in which case $r_\xi < l_\xi$. By the preceding lemma, this implies that $x_i^* = l_i$ for all $i \ne \xi$. But then the inequality $r_\xi < l_\xi$ means that $l_1 + \cdots + l_n > \gamma$, contradicting feasibility. ∎

We have proved that the algorithm returns a feasible point. We have also proved that it returns a point satisfying the conditions of Lemma 5.1. Accordingly, the remaining hurdle is to show that the point satisfies Lemma 5.2. We prove a preliminary lemma, and then the main theorem.

LEMMA 5.6. *Focus on an iteration $k \leq m$, where m is the iteration singled out by Lemma 5.4. Then for any $i \notin F$, $r_i > r_i^-$.*

PROOF. We have already shown that $r_i \geq r_i^-$ in Lemma 5.3; in fact, we showed that the difference between r_i^- and r_i is precisely $u_\xi - x_\xi^*$. Since $k \leq m$, we know that $x_\xi^* = l_\xi$, and therefore $x_\xi^* < u_\xi$. Thus, r_i is greater than r_i^-. ∎

THEOREM 5.4. *Assume that the original problem is feasible. Then the point returned by Algorithm LOCAL-MIN1 is a local minimum.*

PROOF. For the purpose of simplifying notation in the following proof, we assume that the indices $\{1, \ldots, n\}$ are added to F in consecutive order, i.e., x_i^* is determined by the ith iteration of the algorithm. The lemma is proved by showing that the conditions of Lemma 5.2 are satisfied. Consider the value r_m calculated at iteration m singled out by Lemma 5.4.

Let $\lambda_0 = q_m'(r_m^{(m)})$ (we now use a superscript to explicitly indicate the iteration of r_m). As we shall see, we will take λ of Lemma 5.2 to be λ_0 except in the case of degeneracy.

We know $x_i^* = l_i$ for all $i < m$. Focus on a particular $i < m$. Since i rather than m was added to F at iteration i, $q_i'(l_i) \geq q_m'(\max(l_m, r_m^{(i)}))$. Moreover, q_m' is a strictly decreasing function and $r_m^{(m)} > \max(l_m, r_m^{(i)})$ by the preceding lemma. Therefore, $q_m'(\max(l_i, r_m^{(i)})) > q_m'(r_m^{(m)})$. Combining the chain of inequalities shows that $q_i'(l_i) > \lambda_0$.

Now, suppose $r_m^{(m)} < u_m$ (this is the nondegenerate case). In this case take λ in Lemma 5.2 to be λ_0 defined above. We have already shown that $q_i'(x_i^*) > \lambda$ whenever $x_i^* = l_i$. We have also shown that $q_i'(x_i^*) = \lambda$ whenever $l_i < x_i^* < u_i$ (indeed, this particular i is m). Finally, we must show that $q_i'(x_i^*) < \lambda$ whenever $x_i^* = u_i$. If $x_i^* = u_i$ then $i > m$ by our notational assumption. This means that at iteration m of the algorithm, $i \notin F$. Let p_i denote $\max(l_i, r_i^{(m)})$. Since m was added to F rather than i, we know that $q_m'(r_m^{(m)}) \geq q_i'(p_i)$. Moreover, $p_i < u_i$; to see this, note that $l_i < u_i$ and $r_i^{(m)} < u_i$. The latter inequality follows because $u_i - r_i^{(m)} = u_m - r_m^{(m)}$ and we are assuming that $r_m^{(m)} < u_m$. Thus, $p_i < u_i$ and q_i' is a strictly decreasing function. Hence $q_i'(p_i) > q_i'(u_i)$. Combining the inequalities of this paragraph shows that $q_m'(r_m^{(m)}) > q_i'(u_i)$ which is what we needed to prove.

Finally, we must address the degenerate case that $r_m^{(m)} = u_m$. In this case, $x_i^* \in \{l_i, u_i\}$ for all i. We have already shown that if $\lambda_0 = q_m'(r_m^{(m)})$ then $\lambda_0 < q_i'(l_i)$ for $i < m$. We claim also that $\lambda_0 \geq q_i'(u_i)$ for all $i \geq m$;

this follows from arguments given in the previous paragraph. Accordingly, in this case we choose λ for Lemma 5.2 to be $\lambda_0 + \delta$ for a small $\delta > 0$. The conditions of the lemma will be satisfied if δ is small enough. ∎

5.4. An efficient implementation of local minimization

We now turn to the problem of finding an efficient implementation of Algorithm LOCAL-MIN1 from the last section. An implementation of the algorithm exactly as written would require $O(n^3)$ arithmetic operations. The main loop of the algorithm is executed n times. Each iteration requires $O(n)$ values of r_i to be computed. Each r_i requires $O(n)$ steps to compute according to (5.6). Thus, the total product is $O(n^3)$ arithmetic operations.

It is easy to reduce this complexity to $O(n^2)$ steps. In particular, one notices that (5.6) actually requires only $O(1)$ steps to be evaluated if we save the sum of x_j^* for $j \in F$ and the sum of u_j for $j \notin F$. These two totals are each updated with one operation at the beginning of every iteration to reflect the index added to F. From these two totals it is clear that (5.6) can be evaluated with a constant number of arithmetic operations.

The goal of this section is to prove that the complexity can actually be reduced to $O(n \log n)$ steps. In order to prove this, we need to evaluate the quantities of the algorithm in a different order and take advantage of binary search.

In order to describe the binary search, we introduce a new sequence of sets $G^{(k)}$. We define

$$G^{(k)} = \{i : i \notin F^{(k)} \text{ and } r_i^{(k)} > l_i\} \tag{5.8}$$

for $k = 1, \ldots, m$, where m is as in Lemma 5.4. We begin with two lemmas about $G^{(k)}$.

LEMMA 5.7. *The following inclusion holds for k between* 1 *and* $m - 1$ *in Algorithm LOCAL-MIN1:* $G^{(k)} \subset G^{(k+1)}$.

PROOF. Suppose $i \in G^{(k)}$. This means that $i \notin F^{(k)}$ and $r_i^{(k)} > l_i$. But since the variables r_i are nondecreasing as the algorithm progresses, then we know that $r_i^{(k+1)} > l_i$. Moreover, we cannot have $i \in F^{(k+1)}$ since Algorithm LOCAL-MIN1 during its first $m - 1$ iterations can add an index i to F only when $l_i \geq r_i$. Thus, $i \in G^{(k+1)}$. ∎

This lemma tells us that we can simulate the first $m - 1$ iterations of Algorithm LOCAL-MIN1 very efficiently. In particular, a way to reinterpret the first $m - 1$ iterations is as follows. The algorithm seeks $i \notin F$ such that $l_i \geq r_i$ and $q_i'(l_i)$ is largest. This index i is added to F, and the loop is repeated. These steps can be simulated first by sorting the variables so that

$$q_1'(l_1) \geq q_2'(l_2) \geq \cdots \geq q_n'(l_n). \tag{5.9}$$

This sorting requires $O(n \log n)$ steps (see Aho, Hopcroft and Ullman [1974] and Knuth [1973b]). Then the variables can be considered in the order $i = 1, \ldots, n$. For each variable i in this order, we add it to F or discard it because it is in G. Discarding means that the variable is not considered again for membership in F during the first $m - 1$ iterations. The rule for deciding is: We add i to F if $l_i \geq r_i$, else we discard i. We halt the process when F has $m - 1$ elements.

The calculations required by this simulation require $O(n)$ steps, not counting the sorting step. We need to compute r_i for each i, which can be evaluated with a constant number of arithmetic operations using the technique mentioned at the beginning of the section.

The difficulty with this simulation is that we do not know m in advance, and, in fact, it cannot be determined in a simple manner. Accordingly, we run the above simulation until all the variables are used up (either added to F or discarded). This will take us past iteration m of Algorithm LOCAL-MIN1. We now explicitly write down this simulation.

ALGORITHM SIMUL

```
Sort the variables to satisfy (5.9).
Let F^(1) := ∅; k := 1.
For i := 1,...,n
    Compute r_i according to (5.6).
    If l_i ≥ r_i then
        Let x*_i := l_i.
        Let k := k + 1.
        Let F^(k) := F^(k-1) ∪ {i}.
    end if.
end for.
```

We have argued that the sequence of sets $F^{(1)}, F^{(2)}, \ldots, F^{(m)}$ computed by Algorithm SIMUL is the same (assuming ties are broken in the same way) as the sequence computed by Algorithm LOCAL-MIN1. Let k_f denote the final value of k taken by SIMUL.

In order to be consistent with Algorithm LOCAL-MIN1, we define an "iteration" of Algorithm SIMUL to be the operations between successive increments in k. It would be more natural to define an iteration in terms of i rather than k. We see however, that the r_i's and sets F actually depend only on the iteration count k.

Notice that Lemma 5.6 holds for all iterations of Algorithm SIMUL. This is because the difference between $r_i^{(k+1)}$ and $r_i^{(k)}$ is $u_\xi - l_\xi$ for some ξ, which is always positive. Note that even though Algorithm SIMUL does not compute all of the r_i on each iteration, they are still well-defined according to (5.6).

Next, we address the problem of finding m. During iteration m Algorithm LOCAL-MIN1 selects an index $i \in G^{(m)}$ to be in F (because $r_i > l_i$

for the index selected at step m). This means in particular that $G^{(m)}$ is nonempty. The index $i \in G^{(m)}$ chosen by LOCAL-MIN1 is the index such that $q_i'(r_i)$ is maximized. Therefore, for each iteration we define

$$w = \max\{q_i'(r_i) : i \in G\}.$$

This number w depends implicitly on the iteration number, and the dependence will sometimes be explicitly indicated as $w^{(k)}$. The properties of iteration m detected by Algorithm LOCAL-MIN1 and sufficient to prove Theorem 5.4 are:

Condition 1. For all $i \in F$, $w < q_i'(l_i)$.

Condition 2. For all $i \notin F \cup G$, $w \geq q_i'(l_i)$.

It is easy to see that Condition 2 characterizes Algorithm LOCAL-MIN1. This is because if $i \notin F \cup G$, then $q_i'(l_i)$ would be a candidate for the maximum computed by LOCAL-MIN1 during iteration m. Since the actual selection at that step is that $x_\xi^* = r_\xi$ for some $\xi \in G$ such that $w = q_\xi'(r_\xi)$, we know that $w \geq q_i'(l_i)$.

Condition 1 characterizes Algorithm LOCAL-MIN1 because of a chain of inequalities proved earlier. If $i \in F$, then consider the iteration k when i was added to F. Let p denote $\max(l_\xi, r_\xi^{(k)})$. Then because of the selection of i in Algorithm LOCAL-MIN1, we know that $q_i'(l_i) \geq q_\xi'(p)$, and we know $p < r_\xi^{(m)}$, hence $q_\xi'(p) > q_\xi'(r_\xi^{(m)})$. Thus, we conclude that Condition 1 indeed characterizes Algorithm LOCAL-MIN1.

Conditions 1 and 2 are sufficient for the proof of Theorem 5.4. This is the subject of Exercise 5.5. In particular, the conditions of Lemma 5.2 hold.

Now we propose a binary search to find an iteration of Algorithm SIMUL where Conditions 1 and 2 hold. The binary search is done on an interval $[l, u]$ of iteration indices. The invariant property of the binary search is that Condition 1 is satisfied at iteration l but not u. We need a few lemmas to motivate this binary search.

LEMMA 5.8. *Let k_s be the first iteration of Algorithm SIMUL in which an index is discarded (i.e., the condition of the "if" statement in the algorithm is false). Then Condition* 1 *holds at that iteration.*

PROOF. The focus is on iteration k_s. Since G is nonempty at this iteration (in particular, it includes the discarded index), w exists. A chain of inequalities proves the claim. In particular, suppose $w = q_\xi'(r_\xi)$ for some $\xi \in G$. Choose an arbitrary $i \in F$ and suppose i was added to F on iteration $k_0 < k_s$. Then

$$\begin{aligned} w^{(k_s)} &= q_\xi'(r_\xi) \\ &< q_\xi'(l_\xi) \\ &\leq q_i'(l_i). \end{aligned}$$

The first inequality follows because $\xi \in G$. The second inequality comes from the fact that no indices were discarded before iteration k_s, so in particular, ξ must come after i in sorted order (5.9). ∎

LEMMA 5.9. *Suppose $i \notin F \cup G$ and $j \in F^+$. Then $q_i'(l_i) \leq q_j'(l_j)$.*

PROOF. Let the present iteration be k. Since $j \in F^+$, Algorithm SIMUL selected j during one of the iterations between 1 and k. On the other hand, since $i \notin F \cup G$, this means that $r_i \leq l_i$, and i was not discarded or selected SIMUL on iterations 1 to $k - 1$, and was not discarded on iteration k. This means that j comes before i in (5.9), or else $j = i$. ∎

LEMMA 5.10. *Let m be an iteration such that Condition 1 is satisfied at iteration m but not at iteration $m + 1$. Then Conditions 1 and 2 are both satisfied at iteration m.*

PROOF. The focus is on iteration m, so that superscript '+' refers to iteration $m + 1$. Let $\xi \in G^+$ be the index such that $w^+ = q_\xi'(r_\xi^+)$. By assumption, we know that $w^+ \geq q_j'(l_j)$ for some distinguished $j \in F^+$. We take two cases: either $\xi \in G$ or $\xi \notin G$.

If $\xi \in G$ then we have the following chain of inequalities.

$$w \geq q_\xi'(r_\xi) > q_\xi'(r_\xi^+) = w^+ \geq q_j'(l_j).$$

Choose an index $i \notin F \cup G$. Then the previous lemma says that $q_j'(l_j) \geq q_i'(l_i)$. Thus, we have a chain of inequalities showing that $w > q_i'(l_i)$ for all $i \notin F \cup G$, which establishes Condition 2.

The second case is that $\xi \notin G$. We also know that $\xi \notin F$ since $\xi \in G^+$.

We claim in fact that this is a contradiction. In particular, we have the following chain of inequalities. The first line below is derived from the previous lemma since $\xi \notin F \cup G$ and $j \in F^+$:

$$\begin{aligned} q_j'(l_j) &\geq q_\xi'(l_\xi) \\ &> q_\xi'(r_\xi^+) \\ &> w^+. \end{aligned}$$

This contradicts the choice of j made at the beginning of the proof. ∎

Now we can describe the binary search. First, recall that k_f is defined to be the last value of k during Algorithm SIMUL. Notice that

$$F^{(k_f)} \cup G^{(k_f)} = \{1, \ldots, n\}$$

meaning that SIMUL has scanned every variable. If $G^{(k_f)}$ is empty, this means that every variable has been added to F, and in particular, that $l_1 + \cdots + l_n = \gamma$. Therefore, the problem has a unique feasible solution and we are finished.

Otherwise, $G^{(k_f)}$ is nonempty. This means that Condition 2 is vacuously satisfied at iteration k_f since there is no i satisfying the qualifier of the condition. Suppose Condition 1 is also satisfied at iteration k_f; again, we are done since we have found an iteration at which both conditions are satisfied (take $m = k_f$).

Therefore, we are left with the case that $G^{(k_f)}$ is nonempty and Condition 1 is not satisfied at k_f. From Algorithm SIMUL we can identify an iteration k_s in which Condition 1 is satisfied (see Lemma 5.8).

Accordingly, in this case we have identified an interval of iterations $[k_s, k_f]$ such that Condition 1 is satisfied at k_s but not k_f. We repeatedly bisect the interval, maintaining the invariant property that Condition 1 is satisfied at the left end of the interval but not at the right end. Ultimately we reach the point where we have an interval $[m, m+1]$ such that Condition 1 is satisfied at m but not $m+1$. At this point Lemma 5.10 tells us that we have satisfied the conditions at iteration m.

We now derive a bound on the number of operations in this binary search. Since $1 \leq k_s \leq k_f \leq n$, there are at most $\lg n$ phases to the binary search. In each phase, it is necessary to evaluate r_i's, G, and w. From this information it is possible to check Condition 1. All of this information may be computed in $O(n)$ steps. Thus, the binary search described in the last few paragraphs requires $O(n \log n)$ steps total.

Once m is recovered, we can find the optimal point $\mathbf{x}^*$ by noting how Algorithm LOCAL-MIN1 computes $\mathbf{x}^*$. In particular, for all $i \in F^{(m)}$, $x_i^* = l_i$. For the $i \in G^{(m)}$ such that $w^{(m)} = q_i'(r_i)$ (ties broken arbitrarily as usual), we let $x_i^* = r_i$. For all other values of i, we let $x_i^* = u_i$.

We can sum up the operation counts. Sorting requires $O(n \log n)$ operations, SIMUL requires $O(n)$ operations, the binary search for m requires $O(n \log n)$ operations, and the final computation of $\mathbf{x}^*$ requires $O(n)$ operations. This concludes our claim that a local minimum for (5.5) may be found in $O(n \log n)$ steps.

It is interesting to note that we never used the fact that q is a quadratic function; we only need strict concavity and the ability to evaluate q' at selected points. (The only exception is the proofs of Lemma 5.1 and Lemma 5.2, but these proofs can be redone so that they do not depend on the quadratic nature of the objective function.) Therefore, the algorithm of this section could be extended to any strictly concave differentiable separable objective function on $\mathbb{R}^n$. The total running time is then $O(pn \log n)$ where p is the time required to evaluate one partial derivative of the objective function at one particular point.

5.5. The indefinite case of QKP

The indefinite case of QKP is the case that the diagonal entries of D in (5.4) are not restricted in sign. This is the most general case and includes as special cases the convex and concave problems. Since the concave case is **NP**-hard, the same result holds for the indefinite case.

Since we have already demonstrated a polynomial time algorithm for local minima in the concave case, we should explain why we are considering the indefinite case at all. The reason is that, in general, indefinite problems appear to be "harder" than concave problems.

For example, for problems defined on a simplex (see the previous chapter), the convex and concave cases are both solvable in polynomial time. In particular, the convex case is solved with the algorithm of Chapter 3, and the concave case is solved by testing each vertex of the simplex (one can prove that the global minimum of a concave problem always occurs at a vertex; compare this to Lemma 5.1). The indefinite case, however, is **NP**-hard, as proved in Chapter 4.

The result of this section is that a local minimum for the indefinite case of (5.4) can be found in polynomial time. This section is based on Vavasis [1990b]. In that paper a running time of $O(n(\log n)^2)$ is established, but in this book we will prove only an $O(n^3)$ bound.

We first rewrite the problem as follows:

$$\begin{aligned} \text{minimize} \quad & q_1(x_1) + \cdots + q_s(x_s) + p_1(y_1) + \cdots + p_t(y_t) \\ \text{subject to} \quad & x_1 + \cdots + x_s + y_1 + \cdots + y_t = \gamma, \\ & l_i \le x_i \le u_i \quad \text{for } i = 1, \ldots, s, \\ & l'_i \le y_i \le u'_i \quad \text{for } i = 1, \ldots, t. \end{aligned} \tag{5.10}$$

Here, the q_i's are strictly concave quadratic functions of one variable (negative coefficient in the quadratic term) and the p_i's are convex quadratic functions (nonnegative coefficient in the quadratic term). We let $n = s + t$, which is the total number of variables. We assume $s \ge 1$ and $t \ge 1$, because the pure convex and concave cases have already been addressed.

The first step toward solving this problem is to merge the convex variables into one variable. This is done via parameterization. Let $\omega = y_1 + \cdots + y_t$. Then the equation constraint simplifies to $x_1 + \cdots + x_s + \omega = \gamma$. The inequality constraints on y's are dropped and replaced by $l_\omega \le \omega \le u_\omega$, where $l_\omega = l'_1 + \cdots + l'_t$ and $u_\omega = u'_1 + \cdots + u'_t$.

The complicated part is to define an objective function term $\psi(\omega)$. For a fixed choice of ω, we want $\psi(\omega)$ to be equal to $p_1(y_1) + \cdots + p_t(y_t)$ for the "best possible" choices of $y_1, \ldots, y_t$. This motivates the following definition for $\psi(\omega)$.

$$\psi(\omega) = \min \left\{ p_1(y_1) + \cdots + p_t(y_t) : \mathbf{e}^T \mathbf{y} = \omega;\ l'_i \le y_i \le u'_i \right\}. \tag{5.11}$$

Thus, $\psi(\omega)$ is defined to be the optimal value to a convex instance of QKP that is parameterized by ω. With this definition in place, we can consider a new version of the indefinite problem.

$$\begin{aligned} \text{minimize} \quad & q_1(x_1) + \cdots + q_s(x_s) + \psi(\omega) \\ \text{subject to} \quad & x_1 + \cdots + x_s + \omega = \gamma, \\ & l_i \le x_i \le u_i \quad \text{for } i = 1, \ldots, s, \\ & l_\omega \le \omega \le u_\omega. \end{aligned} \tag{5.12}$$

The main reason for creating this new problem is as follows.

LEMMA 5.11. *Let* $(\mathbf{x}^*, \omega^*)$ *be a local minimum for* (5.12). *Let* $\mathbf{y}^*$ *be the value of* $\mathbf{y}$ *in which the minimum in* (5.11) *is achieved for* $\psi(\omega^*)$. *Then* $(\mathbf{x}^*, \mathbf{y}^*)$ *is a local minimum for* (5.10).

PROOF. First, observe that $(\mathbf{x}^*, \mathbf{y}^*)$ is feasible for (5.10); this follows from the fact that $\mathbf{e}^T\mathbf{y}^* = \omega^*$.

Since $(\mathbf{x}^*, \omega^*)$ is a local minimum for (5.12), there is an open set $S \subset \mathbb{R}^{s+1}$ around this point such that if $(\mathbf{x}, \omega) \in S$ and $(\mathbf{x}, \omega)$ is feasible in (5.12), then

$$q_1(x_1) + \cdots + q_s(x_s) + \psi(\omega) \ge q_1(x_1^*) + \cdots + q_s(x_s^*) + \psi(\omega^*).$$

Define $S_0 \subset \mathbb{R}^{s+t}$ to be the inverse image of S under the map

$$(\mathbf{x}, \mathbf{y}) \mapsto (\mathbf{x}, \mathbf{e}^T\mathbf{y}).$$

Observe that S_0 is an open set containing $(\mathbf{x}^*, \mathbf{y}^*)$. The claim is that $(\mathbf{x}^*, \mathbf{y}^*)$ is the minimizer for (5.10) in neighborhood S_0. To see this, consider any other feasible point $(\mathbf{x}, \mathbf{y})$ in S_0. Let $\omega = \mathbf{e}^T\mathbf{y}$; this means that $(\mathbf{x}, \omega) \in S$. The first inequality in the following chain follows from the definition of $\psi(\omega)$; the rest have already been established during this proof.

$$\begin{aligned} & q_1(x_1) + \cdots + q_s(x_s) + \\ & \quad p_1(y_1) + \cdots + p_t(y_t) \ge q_1(x_1) + \cdots + q_s(x_s) + \psi(\omega) \\ & \qquad \ge q_1(x_1^*) + \cdots + q_s(x_s^*) + \psi(\omega^*) \\ & \qquad \ge q_1(x_1^*) + \cdots + q_s(x_s^*) + p_1(y_1^*) + \cdots + p_t(y_t^*). \end{aligned}$$

This chain of inequalities establishes the lemma. ∎

We now explain how to construct $\psi(\omega)$. We apply the HKL algorithm of Chapter 3 to (5.11) and compute the (point-to-set) function $h(\lambda)$. Using sorting, it is possible to completely construct $h(\lambda)$ in $O(t \log t)$ operations. By "completely construct" we mean enumerate all its breakpoints and the coefficients of the linear functions between breakpoints.

The last step of the HKL algorithm is to solve for a particular λ^* such that $h(\lambda^*) = \omega$ (ω is the right hand side of the equation constraint in (5.11)). Since we are treating ω as an indeterminate, we skip this step and instead compute the functional inverse of h as a point-to-set function. Since h is nondecreasing and piecewise linear, its inverse has these same properties. We denote the inverse as $\lambda(\omega)$ since the inverse expresses the dependence of the optimal λ on the varying right-hand side.

This inversion process requires $O(t)$ computations; each segment of h requires a constant number of arithmetic operations to invert.

Next, we recall functions $y_1(\lambda), \ldots, y_t(\lambda)$ that describe the dependence of the optimal y_i on λ. These can be composed with $\lambda(\omega)$ giving functions $\hat{y}_1(\omega), \ldots, \hat{y}_t(\omega)$ indicating the dependence of the optimal point on ω.

In fact, we define $\hat{y}_i$ with the identity $\hat{y}_i(\omega) = y_i(\lambda(\omega))$, with the following exception. Suppose that $y_i(\lambda_1)$ is multiply-valued for an argument λ_1. Suppose $\lambda_1 = \lambda(\omega_1)$ for an argument ω_1. Then we define $\hat{y}_i(\omega_1)$ to have the single value chosen so that

$$\hat{y}_1(\omega_1) + \cdots + \hat{y}_t(\omega_1) = \omega_1.$$

Finally, we can write a formula for $\psi(\omega)$:

$$\psi(\omega) = p_1(\hat{y}_1(\omega)) + \cdots + p_t(\hat{y}_t(\omega)). \tag{5.13}$$

Between breakpoints, λ varies linearly with ω, y_i varies linearly with λ, and p_i varies quadratically with y_i. Between a particular pair of breakpoints the coefficients of the quadratic function for $\psi(\omega)$ in terms of ω can be computed in $O(t)$ operations by summing up the contributions from each term. Therefore, the time to compute the entire graph of $\psi(\omega)$ given the graph of $\lambda(\omega)$ is $O(t^2)$. In fact, this can be reduced to $O(t)$ by noting that the coefficients can be updated between consecutive breakpoints rather than recomputed from scratch.

Thus, $\psi(\omega)$ defined by (5.11) is a piecewise quadratic function that can be completely computed in $O(t \log t)$ steps, which is bounded by $O(n \log n)$ steps. This function has the additional important property of convexity. This is the content of the following lemma, which is stated more generally with no additional complication.

LEMMA 5.12. *Let $Y \subset \mathbb{R}^t$ be a convex compact region. Suppose $z \in \mathbb{R}$ and $\mathbf{b} \in \mathbb{R}^t$ are fixed, and suppose $\chi : Y \to \mathbb{R}$ is a continuous convex function. Define the real-valued function ψ by $\psi(\omega) = \min\{\chi(\mathbf{y}) : \mathbf{y} \in Y, \mathbf{b}^T\mathbf{y} = \omega + z\}$. Then ψ is defined on a closed interval, and in addition, ψ is a convex function on the interval.*

PROOF. The region on which ψ is defined is the image of Y under the map $\mathbf{y} \mapsto \mathbf{b}^T\mathbf{y} - z$; since Y is convex and compact, this image must be a closed interval. Let us consider two values of the parameter ω' and ω'' lying in this interval. Let λ be an arbitrary number in the interval $[0, 1]$. Note that $\psi(\omega')$ is defined as the minimum of continuous function on a compact set, hence the minimum is achieved, say at $\mathbf{y}'$. Similarly $\psi(\omega'') = \chi(\mathbf{y}'')$ for some feasible $\mathbf{y}''$. Thus,

$$\begin{aligned}\lambda\psi(\omega') + (1-\lambda)\psi(\omega'') &= \lambda\chi(\mathbf{y}') + (1-\lambda)\chi(\mathbf{y}'') \\ &\geq \chi(\lambda\mathbf{y}' + (1-\lambda)\mathbf{y}'').\end{aligned}$$

The second line above follows from the convexity of χ.

By choice of $\mathbf{y}'$ and $\mathbf{y}''$ we know that

$$\begin{aligned}\mathbf{b}^T\mathbf{y}' &= \omega' + z, \\ \mathbf{b}^T\mathbf{y}'' &= \omega'' + z, \text{ and} \\ \mathbf{y}', \mathbf{y}'' &\in Y.\end{aligned}$$

Let $\mathbf{y} = \lambda\mathbf{y}' + (1-\lambda)\mathbf{y}''$. Then notice that $\mathbf{y} \in Y$ because $\mathbf{y}$ is a convex combination of two points already in Y. Next, let $\omega = \mathbf{b}^T\mathbf{y} - z$. Then

$$\begin{aligned}\chi(\lambda\mathbf{y}' + (1-\lambda)\mathbf{y}'') &= \chi(\mathbf{y}) \\ &\geq \psi(\omega).\end{aligned}$$

The first line follows from the definition of $\mathbf{y}$ and the second from the fact that $\psi(\omega)$ is defined to be the minimum value of $\chi(\mathbf{y})$ over all $\mathbf{y} \in Y$ that satisfy $\mathbf{b}^T\mathbf{y} = \omega + z$.

The two preceding chains of inequalities show that

$$\lambda\psi(\omega') + (1-\lambda)\psi(\omega'') \geq \psi(\omega).$$

Tracing back the definition of ω will show that $\omega = \lambda\omega' + (1-\lambda)\omega''$. Thus, the preceding inequality establishes the convexity of ψ. ∎

The function $\psi(\omega)$ defined by (5.11) is convex, as we have seen, but in general will not be differentiable. Nonetheless, we will want to talk about the derivative of ψ, so we introduce the notion of the subdifferential of ψ. Let us assume that ψ is defined on $\mathbb{R}$ (which will be the case subsequently). Then we regard $\psi'(z)$ for $z \in \mathbb{R}$ as a set rather than as a single number. In particular, we say that $\zeta \in \psi'(z)$ for some z iff

$$\psi(z+d) \geq \psi(z) + d\zeta$$

for all $d \in \mathbb{R}$. With this definition, the set $\psi'(x)$ is called the *subdifferential* of ψ at x. Note that this definition for a general continuous function ψ would typically yield the empty set. It can be proved, however, that because ψ is convex, this set is a nonempty closed interval for all z. Indeed, this interval will be $[\psi'_-(z), \psi'_+(z)]$, where ψ'_-, ψ'_+ denote left and right derivatives. At a point where ψ is differentiable, this definition will imply that ψ' is the singleton set whose element is the ordinary derivative. See Fletcher [1987] for more information.

Interestingly, it turns out that for the problem under consideration, the identity $\psi'(\omega) = \lambda(\omega)$ holds. This can be verified from (5.13). This identity does not appear to lead to asymptotic running time improvements.

Next we want to show that we can redefine $\psi(\omega)$ so that it is finite, continuous, and convex for all real ω. So far we have assumed that $\psi(\omega) = \infty$ if (5.11) is infeasible—in particular, so far $\psi(\omega)$ is finite only for

$$l_\omega \leq \omega \leq u_\omega$$

because this is when the system implicit in (5.11) is feasible. Define a new objective function $\hat{\psi}(\omega)$ for all of $\mathbb{R}$. On $[l_\omega, u_\omega]$, define $\hat{\psi} = \psi$. Let θ be a very large number. Set $\hat{\psi}(\omega) = \theta\omega + \alpha_1$ for $\omega > u_\omega$ and $\hat{\psi}(\omega) = -\theta\omega + \alpha_2$ for $\omega < l_\omega$. We choose α_1, α_2 to make $\hat{\psi}$ continuous at l_ω and u_ω. Thus, we have the new system

$$\begin{aligned} \text{minimize:} \quad & q_1(x_1) + \cdots + q_s(x_s) + \hat{\psi}(\omega) \\ \text{subject to:} \quad & x_1 + x_2 + \cdots + x_s + \omega = \gamma, \\ & l_i \leq x_i \leq u_i, \quad i = 1, \ldots, s. \end{aligned} \tag{5.14}$$

How large does θ have to be? One requirement is that it must be large enough so that $\hat{\psi}$ maintains its convexity, i.e., larger than $|\psi'(\omega)|$ for any $\omega \in [l_\omega, u_\omega]$. We also require θ to be larger in magnitude than the derivative of any $q_i(x_i)$ for any $x_i \in [l_i, u_i]$. With these restrictions, the following lemma tells us that (5.14) is equivalent to (5.12).

LEMMA 5.13. *If the original system* (5.12) *is feasible, then any local minimum for* (5.14) *satisfies* $\omega \in [l_\omega, u_\omega]$.

PROOF. Let $(\mathbf{x}^*, \omega^*)$ be a local minimum for (5.14). We have to show that $\omega^* \in [l_\omega, u_\omega]$. Suppose not, e.g., suppose $\omega^* > u_\omega$. Since the original system was feasible, there is some i such that $x_i^* < u_i$. Then we claim that for any $\epsilon \in (0, \min(\omega^* - u_\omega, u_i - x_i^*))$, the feasible point

$$(x_1^*, \ldots, x_{i-1}^*, x_i^* + \epsilon, x_{i+1}^*, \ldots, x_s^*, \omega^* - \epsilon)$$

has a smaller value of the objective function than the supposed local minimum. This follows because $\hat{\psi}(\omega^* - \epsilon) = \hat{\psi}(\omega^*) - \theta\epsilon$, whereas $q_i(x_i^* + \epsilon) < q_i(x_i^*) + \theta\epsilon$ by the mean value theorem and our assumption that $\theta > |q_i'(x)|$. This contradicts the local minimality of $(\mathbf{x}^*, \omega^*)$. Similarly, if we assume that $\omega^* < l_\omega$, this also leads to a contradiction. ∎

From now on we work on (5.14) rather than (5.12), although we write ψ rather than $\hat{\psi}$. Thus, we can state:

ASSUMPTION 5.1. *The function* ψ *in* (5.14) *is convex, finite, and continuous for all real* ω.

We can also make the following assumption as we did in the last two sections.

ASSUMPTION 5.2. *For each* i, $l_i < u_i$.

We now try to characterize a local minimum. Lemma 5.1 applies immediately to the problem at hand with the same proof. In other words, one property of a local minimum is that at most one of $(x_1^*, \ldots, x_s^*)$ is away from boundary values. As a result of the transformation described above, ω does not have boundary values and is not restricted by the lemma. We now would like to investigate the analog of Lemma 5.2.

According to the argument in the previous paragraph, there are two types of local minima: those with all of the x_i's at boundary values, and those with all but one of the x_i's at boundary values. We note that neither of these cases is degenerate.

The next task is to characterize local minima of the two types. This is the purpose of the upcoming two lemmas. Let

$$\boldsymbol{\pi}^* = (x_1^*, \ldots, x_s^*, \omega^*)$$

be a feasible point. We have just seen a result that implies that all of the numbers $x_1^*, \ldots, x_s^*$, except for possibly one, are at boundary values if this point is a local minimum. Let F be the set of i such that $x_i^* = l_i$.

LEMMA 5.14. *Let* $\overline{F}$ *denote* $\{1, \ldots, s\} - F$. *Suppose that for all* $i \in \overline{F}$, $x_i^* = u_i$. *Then* $\boldsymbol{\pi}^*$ *is a local minimum of* (5.14) *if and only if there exists an* $\epsilon > 0$ *such that*

(i) $q_i'(x_i^*) > q_j'(x_j^*)$ *for all* $i \in F$, $j \in \overline{F}$,

(ii) $q_i'(x_i^* + \delta) \geq \psi'(\omega^* - \delta)$ *for all* $\delta \in (0, \epsilon)$ *and for all* $i \in F$, *and*

(iii) $q_i'(x_i^* - \delta) \leq \psi'(\omega^* + \delta)$ *for all* $\delta \in (0, \epsilon)$ *and for all* $i \in \overline{F}$.

A local minimum satisfying these conditions will be called a **Type A** *local minimum.*

REMARK. We can compare these conditions to the KKT conditions: They are similar, except we require that our inequalities hold in intervals rather than at an individual point alone.

PROOF. Let us write the objective function as $f(\boldsymbol{\pi})$. Note that the problem under consideration is not a quadratic program, because ψ is not a quadratic function. There exists an $\epsilon > 0$, however, such that the problem is quadratic in each of the regions $R_1 = \mathbb{R}^s \times [\omega^*, \omega^* + \epsilon]$ and $R_2 = \mathbb{R}^s \times [\omega^* - \epsilon, \omega^*]$. This is because ψ is piecewise quadratic. Thus, we can prove local minimality by showing that $\boldsymbol{\pi}^*$ is locally optimal in R_1 and in R_2 separately. Let us focus on R_1, since R_2 is handled analogously.

We need to show that Conditions 1 and 2 of Theorem 5.3 are equivalent to Conditions (i)–(iii) of this lemma. Let θ be the right-derivative of ψ at ω^*.

Then the gradient of f at $\mathbf{x}^*$ with respect to R_1 is $\mathbf{g} = (q_1'(x_1^*), \ldots, q_s'(x_s^*), \theta)^T$. Let Z be the Hessian of f at $\boldsymbol{\pi}^*$ with respect to R_1. Then Z is a diagonal matrix whose first s diagonal entries $z_{ii}, i = 1, \ldots, s$ are negative, and whose entry $z_{s+1,s+1}$ is nonnegative.

It is now straightforward to establish the necessity of (i)–(iii). For example, if (iii) were violated for a particular i, then either $g_i > \theta$ (here, g_i refers to the ith entry of vector $\mathbf{g}$ in the last paragraph), or $g_i = \theta$ and $-z_{ii} > z_{s+1,s+1}$. In either case there is an improvement in the objective function by taking $\mathbf{d}$ to be the vector with -1 in position i, 1 in position $s+1$, 0's elsewhere, and $\epsilon > 0$ small enough.

To establish sufficiency, suppose (i)–(iii) hold and let $\mathbf{d} \neq \mathbf{0}$ be a feasible direction with $d_{s+1} \geq 0$. This means that $\mathbf{e}^T\mathbf{d} = 0$, where $\mathbf{e}$ is vector of all 1's. It also means that $d_i \geq 0$ for $i \in F$ and $d_i \leq 0$ for $i \in \overline{F}$. Let $\mu = \max\{q_i'(u_i) : i \in \overline{F}\}$.

Since $\mathbf{e}^T\mathbf{d} = 0$, then $\mathbf{g}^T\mathbf{d} = (\mathbf{g} - \mu\mathbf{e})^T\mathbf{d}$. We claim that each term in the inner product $(\mathbf{g} - \mu\mathbf{e})^T\mathbf{d}$ is nonnegative, which establishes Condition 1 of Theorem 5.3. The ith term is $(g_i - \mu)d_i$. If, for example, $d_i < 0$ then $i \in \overline{F}$. This means that $g_i \leq \mu$ by definition of μ. If $d_i > 0$ then $i \in F \cup \{s\}$. For $i \in F$, $g_i > \mu$ by (i) of the lemma so the inner product term is positive. Similarly, $\theta \geq \mu$ by (iii).

Thus, we are left with the case that $(\mathbf{g} - \mu\mathbf{e})^T\mathbf{d} = 0$, in which case we have to establish that Condition 2 of Theorem 5.3 holds. It cannot be the case that a component of $\mathbf{d}$ indexed by F is nonzero, because (i) of the lemma would imply that one of the terms of the inner product is positive. Thus, the remaining case is that $\mathbf{d}$ is nonzero only in components indexed by $\overline{F} \cup \{s+1\}$.

Accordingly, we look into the case that $\mathbf{d}$ has negative components in positions indexed by some set $J \subset \overline{F}$ and a positive component in position $s+1$ with the same magnitude as the sum of the negative entries. Since we are assuming $\mathbf{g}^T\mathbf{d} = 0$, we conclude that $g_i = \mu = \theta$ for $i \in J$. Then (iii) implies that $-z_{ii} \leq z_{s+1,s+1}$. We claim this means that the second-order term $\mathbf{d}^T Z\mathbf{d}$ is nonnegative. In particular,

$$\mathbf{d}^T Z\mathbf{d} = \sum_{i \in J} d_i^2 z_{ii} + \left(\sum_{i \in J} d_i\right)^2 \cdot z_{s+1,s+1}$$

which is nonnegative because the first term on the right-hand side is dominated by the second. ∎

LEMMA 5.15. *Suppose all of $x_1^*, \ldots, x_s^*$ are at boundary values except for a distinguished $x_p^* \in (l_p, u_p)$. As above, let $F = \{i : x_i^* = l_i\}$. Let $\overline{F}$ denote $\{1, \ldots, s\} - F - \{p\}$ (so that $x_i^* = u_i$ for $i \in \overline{F}$). Then $\boldsymbol{\pi}^*$ is a local minimum of (5.14) if and only if there exists an $\epsilon > 0$ such that*

(i) *$q_i'(x_i^*) > q_p'(x_p^*)$ for all $i \in F$,*

(ii) $q'_p(x^*_p) > q'_j(x^*_j)$ *for all* $j \in \overline{F}$,

(iii) $q'_p(x^*_p + \delta) \geq \psi'(\omega^* - \delta)$ *for all* $\delta \in (0, \epsilon)$, *and*

(iv) $q'_p(x^*_p - \delta) \leq \psi'(\omega^* + \delta)$ *for all* $\delta \in (0, \epsilon)$.

A local minimum satisfying these conditions will be called a **Type B** *local minimum.*

REMARK. If ψ' is single-valued at ω^*, then conditions (iii)–(iv) imply that $\psi'(\omega^*) = q'_p(x^*_p)$.

PROOF. The proof here is completely analogous to the proof of Lemma 5.14, so we omit it. We make the observation that (i)–(ii) imply that $q'_i(x^*_i) > q'_j(x^*_j)$ for all $i \in F$, $j \in \overline{F}$. Also, observe that (i) and (iii) together imply that $q'_i(x^*_i + \delta) > \psi'(\omega^* - \delta)$ for $\delta > 0$ sufficiently small and $i \in F$, and similarly, $q'_j(x^*_j - \delta) < \psi'(\omega^* + \delta)$ for $j \in \overline{F}$. ∎

We remark that conditions like (iii) and (iv) in the preceding lemma would in general be impossible to verify computationally because they involve checking a derivative at a continuum of points. For (5.14) however, we can verify conditions like this with a finite number of operations because the derivatives of the q_i's are linear and the derivative of ψ is piecewise linear.

We now propose an algorithm for finding a local minimum of (5.14). The following algorithm is analogous to the first m steps of Algorithm LOCAL-MIN1. The algorithm is more complicated because we have two variables that could be off-boundary, namely x^*_p as in Lemma 5.15 and ω^*. In addition, the following problem occurs with the technique of Algorithm LOCAL-MIN1. In Algorithm LOCAL-MIN1, as we assigned a variable x^*_i to its lower bound l_i, the value r_ξ for ξ assigned off its boundary value repeatedly increased, which means that $q'_\xi(r_\xi)$ decreased. This means that if $q'_i(l_i) > q'_\xi(r_\xi)$ holds at an iteration, then it holds at future iterations. This type of inequality was instrumental in proving that LOCAL-MIN1 returned a local minimum.

For the current problem, if we assign x^*_i to l_i, then the minimum feasible value for ω increases. This forces $\psi'(\omega)$ to increase, which may cause the conditions for local minimality to be violated.

We solve this problem by introducing new variables $\rho_1, \ldots, \rho_s$. The quantity ρ_i is the smallest value of x_i such that condition (iii) of Lemma 5.15 is satisfied for $\omega = r_i - x_i$. Such a ρ_i may not exist.

The algorithm is as follows.

ALGORITHM LOCAL-MIN2

Let $F^{(1)} := \emptyset$.

for $k := 1, 2, \ldots$

For each $i \notin F^{(k)}$, compute r_i according to (5.6).

For each $i \notin F^{(k)}$, let ρ_i be the smallest number in $[l_i, u_i]$

satisfying (5.17) below and either (5.18) or (5.19).
Let $H := \{i \notin F : \rho_i \text{ does not exist}\}$.
Halt if $F \cup H = \{1, \ldots, s\}$ (call this HALT A).
Let $G := \{i \notin F \cup H : \rho_i > l_i\}$.
Let $w := \max\{q_i'(\rho_i) : i \in G\}$.
If w exists (i.e., $G \neq \emptyset$) and

$$w < q_j'(l_j) \quad \text{for all } j \in F, \text{ and} \tag{5.15}$$

$$w > q_j'(u_j) \quad \text{for all } j \notin F \cup G. \tag{5.16}$$

then halt. (HALT B)
If $F \cup G \cup H = \{1, \ldots, s\}$ then halt. (HALT C.)
Find $i \notin F \cup G \cup H$ such that $q_i'(l_i)$ is maximum.
Let $x_i^* := l_i$.
Let $F^{(k+1)} := F^{(k)} \cup \{i\}$.
end for.

The three conditions referred to above are

$$q_i'(\rho_i) \geq \psi'(r_i - \rho_i), \tag{5.17}$$

$$\rho_i = u_i, \tag{5.18}$$

$$\exists \epsilon > 0 \text{ such that } q_i'(\rho_i + \delta) \geq \psi'(r_i - \rho_i - \delta) \text{ for all } \delta \in (0, \epsilon). \tag{5.19}$$

We remark that this algorithm only assigns $x_i^* = l_i$ and therefore will usually terminate before all the variables have been assigned. In this sense it is like the first $m - 1$ iterations of Algorithm LOCAL-MIN1. Below we explain how to assign the rest of the variables after termination. As we shall see below, HALT A corresponds to a local minimum of Type A, and HALT B corresponds to a local minimum of Type B (or possibly Type A if there is degeneracy). The algorithm will never reach HALT C, as we shall prove.

First, we need to explain some features of the algorithm. In the statement where $i \notin F \cup G \cup H$ is chosen to maximize $q_i'(l_i)$, we break ties by selecting i with the largest value of q_i'' (the function $q_i''(x)$ is constant-valued so we omit its argument; its value is always negative).

In (5.17) it is possible for ψ' to be multiply valued at $r_i - \rho_i$; in this case we say that inequality (5.17) is true if it holds for at least one value of $\psi'(r_i - \rho_i)$.

Now, we explain how to compute ρ_i satisfying the above conditions. Recall that ψ' is a piecewise linear increasing function, all of whose coefficients are computed in advance. The procedure for computing ρ_i is as follows. We first check whether $\rho_i = l_i$ satisfies (5.17) and (5.19). Statement (5.19) is tested by comparing the slope of q_i' at l_i to the slope of ψ' at $r_i - l_i$. If l_i does not satisfy the conditions, then we loop over the breakpoints of ψ', starting with the pair of breakpoints enclosing $r_i - l_i$. Between every pair of

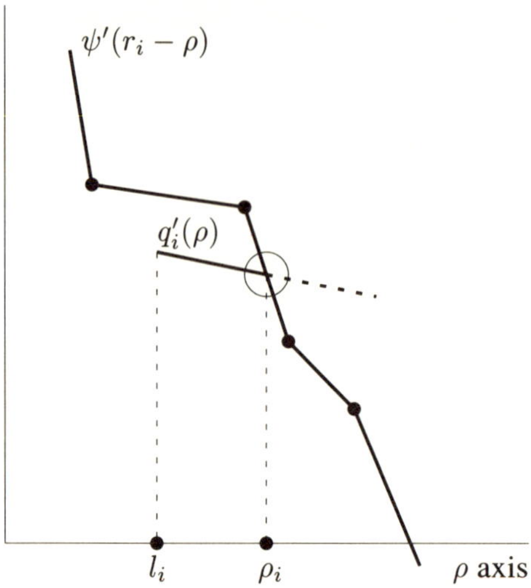

FIG. 5.2. Computation of ρ_i.

breakpoints $\psi'(r_i - \rho)$ varies linearly with ρ, so the potentially smallest value of ρ to satisfy (5.17) can be computed by intersecting two line segments. Similarly, (5.18) and (5.19) are easy to check. Fig. 5.2 shows an example of the determination of ρ_i. We first make the following observations about Algorithm LOCAL-MIN2 that follow directly from the definitions.

1. There exists a unique subscript i such that $F^+ = F \cup \{i\}$.

2. The sets F, G, H of subscripts are always mutually disjoint.

3. The algorithm will terminate in at most s steps with either HALT A, HALT B or HALT C.

Now we begin the series of lemmas to establish properties of this algorithm. We know for the same reason as in the concave case that $r_i > r_i^-$ if $i \notin F$ because $r_i - r_i^- = u_\xi - l_\xi$ which is positive.

LEMMA 5.16. *If $i \notin F$ and ρ_i exists, then ρ_i^- exists and $\rho_i \geq \rho_i^-$. Moreover, $H^- \subset H$.*

PROOF. If ρ_i exists, then the same value of ρ_i will satisfy (5.17) at the previous iteration. This is because r_i is increasing and ψ' is an increasing function of its argument. Moreover, whichever of (5.18) or (5.19) is satisfied by ρ_i will be satisfied by this same value of ρ_i at the previous iteration.

Therefore, ρ_i satisfies all the conditions that ρ_i^- is expected to satisfy. Since ρ_i^- is the minimum value satisfying these conditions, the first statement of the lemma follows.

Now, we turn to the second statement. Suppose $i \in H^-$. Then $i \notin F^-$ and ρ_i^- fails to exist. By the previous part of the lemma, this means ρ_i fails to exist. Moreover, we cannot have $i \in F$ because an index i is added to F only if it is not in H^-. ∎

LEMMA 5.17. *The sequence of $q_j'(l_j)$'s chosen by the algorithm are numerically nondecreasing.*

PROOF. This would follow if we could show that the set $F \cup G \cup H$ is increasing; then every index that is a candidate for the maximum at iteration k would have also been a candidate at iteration $k-1$. First, note that F is increasing. Second, we claim that $G \cup H$ is increasing. We know from the previous lemma that H by itself is increasing. Once an index joins G, it remains for the rest of the algorithm in $G \cup H$. This is because ρ_i either increases or vanishes, but can never decrease. ∎

The following lemma is analogous to Lemma 5.9, and the proof is omitted.

LEMMA 5.18. *Suppose $i \notin F \cup G \cup H$ and $j \in F^+$. Then $q_i'(l_i) \leq q_j'(l_j)$.*

Next we turn to the question of the halting conditions. We want to establish that either HALT A or HALT B is eventually satisfied. This proof is via a series of lemmas.

LEMMA 5.19. *If $G^- = \emptyset$ but $G \neq \emptyset$ then* (5.15) *is satisfied during the current iteration. The same conclusion holds if $G \neq \emptyset$ when $k = 1$.*

PROOF. If $k = 1$ and $G \neq \emptyset$ then (5.15) is vacuously satisfied. For $k > 1$ let i be a typical element of G; we have to show that $q_i'(\rho_i) < q_j'(l_j)$ for all $j \in F$. Since $i \notin G^-$, then $\rho_i^- = l_i$ at that step, i.e., $i \notin F^- \cup G^- \cup H^-$. Then by Lemma 5.18 we know that $q_i'(l_i) \leq q_j'(l_j)$ for all $j \in F$. In addition, $q_i'(l_i) > q_i'(\rho_i)$ by strict concavity. Combining these two establishes the inequality we need to prove. ∎

LEMMA 5.20. *Suppose that $i \notin F \cup H$ and $j \in H$ for subscripts i, j. Then $q_i'(\rho_i) > q_j'(u_j)$.*

PROOF. Since ρ_j does not exist, this means that u_j does not satisfy (5.17), i.e., $q_j'(u_j) < \psi'(r_j - u_j)$. (If $\psi'(r_j - u_j)$ is multiply-valued, then the negation of (5.17) stipulates that $q_j'(u_j) < z$ for all $z \in \psi'(r_j - u_j)$.) Recall that $r_i - u_i$ does not depend on i, so $r_j - u_j = r_i - u_i \leq r_i - \rho_i$, so $\psi'(r_j - u_j) \leq \psi'(r_i - \rho_i)$ by convexity. Finally, $\psi'(r_i - \rho_i) \leq q_i'(\rho_i)$ because (5.17) must hold in the case of i. Combining all of these gives the result that $q_i'(\rho_i) > q_j'(u_j)$. ∎

LEMMA 5.21. *Suppose the algorithm reaches the point that*

$$F \cup G \cup H = \{1, \ldots, s\}. \tag{5.20}$$

Then either condition HALT A holds, or G is nonempty and (5.16) *holds.*

PROOF. Suppose (5.20) holds. There are two cases: either $G = \emptyset$ or $G \neq \emptyset$. If $G = \emptyset$ then (5.20) is equivalent to HALT A so the lemma is true. The other case is that G is nonempty and hence includes all the $i \notin F$ for which ρ_i exists. Thus, for (5.16) we have to show that $w > q_j'(u_j)$ for all $j \in H$. Let i be a typical element of G, so that ρ_i exists. Since w is defined as a max, it is enough to show that $q_i'(\rho_i) > q_j'(u_j)$. But this follows immediately from the preceding lemma. ∎

The following lemma is analogous to Lemma 5.10, but more complicated because there are more ways for the algorithm to halt.

LEMMA 5.22. *Suppose $G \neq \emptyset$, $G^+ \neq \emptyset$, and* (5.15) *holds at the current iteration but not during the next iteration. Then* (5.16) *holds at the current iteration.*

PROOF. We want to show that (5.16) holds, so let $\xi \in G$ be the index such that $q_\xi'(\rho_\xi) = w$. Focus on a particular $i \notin F \cup G$. There are two cases of (5.16), namely, $i \in H$ or $i \notin H$. If $i \in H$ then $q_\xi'(\rho_\xi) > q_i'(u_i)$ because of Lemma 5.20. This inequality establishes (5.16).

Otherwise, if $i \notin H$, then Lemma 5.18 tells us that for any $j \in F^+$, $q_j'(l_j) \geq q_i'(l_i)$. Since (5.15) is assumed not to hold on the next iteration, this implies that there exists a distinguished $m \in G^+$ such that $q_m'(\rho_m^+) \geq q_j'(l_j)$ for some $j \in F^+$.

Next, observe by concavity that $q_m'(\rho_m) \geq q_m'(\rho_m^+)$. Now, there are two cases, either $m \in G$ or $m \notin G$. If $m \in G$, then $q_\xi'(\rho_\xi) \geq q_m'(\rho_m)$ by choice of ξ. Combining the entire chain of inequalities in the last few paragraphs, we conclude that $q_\xi'(\rho_\xi) \geq q_i'(l_i)$, and hence that $q_\xi'(\rho_\xi) > q_i'(u_i)$ by concavity of q_i. This inequality establishes (5.16).

The final case is that $m \notin G$; this case leads to a contradiction. If $m \notin G$, then $m \notin F \cup G \cup H$ (we know that $m \notin F \cup H$ since $m \in G^+$). This implies by Lemma 5.18 that $q_j'(l_j) \geq q_m'(l_m) > q_m'(\rho_m^+)$ for all $j \in F^+$. On the other hand, we stated above that $q_m'(\rho_m^+) \geq q_j'(l_j)$ for some $j \in F^+$. ∎

THEOREM 5.5. *Either HALT A or HALT B is satisfied at some point in the algorithm.*

PROOF. We would hit HALT C on the iteration when (5.20) holds—call this iteration k_f. Suppose that neither HALT A nor HALT B is satisfied at iteration k_f—we demonstrate that one or the other must have held at an earlier iteration. From Lemma 5.21, one possibility is that HALT A holds at k_f, in which case we are done. The other possibility is that G is nonempty and (5.16) holds. Then we back up to the highest-numbered iteration before k_f such that G is nonempty and G^- is empty, say iteration k_g. Focusing on this iteration, we know that (5.15) is satisfied at k_g by Lemma 5.19. By assumption, G is nonempty for all steps between k_g and k_f. Now we look at the first step after k_g at which (5.15) fails. If (5.15) does not fail at all between k_g and k_f then it is still true at k_f in which case both (5.15) and

(5.16) are true at k_f. Otherwise, if (5.15) first fails at $k_e > k_g$, then by the previous lemma both (5.15) and (5.16) are true at iteration $k_e - 1$. ∎

Now we explain how to produce a local minimum if HALT A or HALT B is satisfied.

Let us suppose that HALT A is satisfied at some point in the algorithm; say it is satisfied at iteration m. For the next few paragraphs, the focus is on iteration m. Define $\boldsymbol{\pi}^* = (x_1^*, \ldots, x_s^*, \omega^*)$ according to the rules

$$\begin{aligned} x_i^* &= l_i \quad \text{for } i \in F, \\ x_i^* &= u_i \quad \text{for } i \notin F, \\ \omega^* &= \gamma - x_1^* - \cdots - x_s^*. \end{aligned} \tag{5.21}$$

THEOREM 5.6. *When halting condition HALT A is satisfied, the point* $\boldsymbol{\pi}^*$ *defined by* (5.21) *is a Type A local minimum in the sense of Lemma* 5.14.

PROOF. Proving local minimality reduces to verifying the three conditions of Lemma 5.14. We start with condition (iii). Suppose $i \in \overline{F}$; this implies that $i \in H$ by condition HALT A. Therefore, ρ_i does not exist, hence $q_i'(u_i) < \psi'(r_i - u_i)$, and this holds for all possible such values if ψ' is multiply-valued. But checking definitions shows that $r_i - u_i = \omega^*$, hence $q_i'(x_i^*) < \psi'(\omega^*)$. This implies that condition (iii) holds for a small enough ϵ.

Next, we examine condition (ii). Note that condition (ii) is trivial if $F = \emptyset$, hence assume $F \neq \emptyset$ so that the halting iteration index, m, is at least 2. We look at iteration $m-1$. Let ξ be the index such that $F = F^- \cup \{\xi\}$. Then we know that ρ_ξ^- exists and satisfies $\rho_\xi^- = l_\xi$. This means also by (5.17) and (5.19) that $q_\xi'(l_\xi + \delta) \geq \psi'(r_\xi^- - l_\xi - \delta)$ for all δ small enough. But checking definitions, $r_\xi^- - l_\xi = \omega^*$. Hence condition (ii) is satisfied for this particular value of ξ. Consider any other $i \in F$. We know that $q_i'(l_i) \geq q_\xi'(l_\xi)$ by Lemma 5.17. Moreover, by the tie-breaking for the selection of $i \notin F \cup G \cup H$, we know that $q_i'(l_i + \delta) \geq q_\xi'(l_\xi + \delta)$ for suitably small $\delta > 0$. Therefore, $q_i'(l_i + \delta) \geq \psi'(\omega^* - \delta)$ for all $i \in F$.

Now it is easy to see that condition (i) is satisfied; let z be the smallest value of $\psi'(\omega^*)$. We have proved that $q_i'(l_i) \geq z$ for all $i \in F$ and that $q_j'(u_j) < z$ for all $j \in \overline{F}$. This establishes (i). ∎

This takes care of the case when HALT A is satisfied. Let us turn to the more complicated case, namely halting condition HALT B. Say (5.15) and (5.16) are satisfied at iteration m. For the next few paragraphs, the focus is on iteration m. Let p be the index such that $p \in G$ and $w = q_p'(\rho_p)$. If there are several possible choices for p in this situation, then choose p so that $p \in G$, $w = q_p'(\rho_p)$ and $u_p - \rho_p$ is minimized.

Then define $\boldsymbol{\pi}^* = (x_1^*, \ldots, x_s^*, \omega^*)$ according to the rules

$$\begin{aligned} x_i^* &= l_i \quad \text{for } i \in F, \\ x_p^* &= \rho_p, \end{aligned}$$

$$
\begin{aligned}
x_i^* &= u_i \quad \text{for } i \notin F \cup \{p\}, \\
\omega^* &= \gamma - x_1^* - \cdots - x_s^*.
\end{aligned} \tag{5.22}
$$

THEOREM 5.7. *The point defined by* (5.22) *when* (5.15) *and* (5.16) *are satisfied is either a Type B local minimum in the sense of Lemma* 5.15, *or, in the case that* $\rho_p = u_p$, *a Type A local minimum.*

PROOF. Proving this amounts to verifying the four conditions of the lemma. We assume for now that $\rho_p < u_p$; the other case is treated below.

We start with verifying condition (iii) of Lemma 5.15. It follows from the definitions that $\omega^* = r_p - \rho_p$, and therefore (5.19) is identical to condition (iii) at π^*. Next we turn to (iv). This is true because ρ_p is the minimum value such that (5.19) is true and $\rho_p > l_p$, hence there exists an infinite sequence of positive numbers $\epsilon_1, \epsilon_2, \ldots$ converging to zero such that $q_p'(\rho_p - \epsilon_i) < \psi'(\omega^* + \epsilon_i)$ which implies by piecewise continuity that (iv) holds.

Next, observe that the condition (5.15) is exactly equivalent to condition (i). Finally, we look into condition (ii). Observe that (5.16) implies that $q_p'(\rho_p) > q_i'(u_i)$ for all $i \notin F \cup G$. Thus, it remains to show that

$$
q_p'(\rho_p) > q_i'(u_i)
$$

for all $i \in G$. First, observe that since $q_i'(\rho_i)$ for $i \in G$ is maximized for subscript p, we can claim that $q_p'(\rho_p) \geq q_i'(\rho_i)$. Second, $q_i'(\rho_i) \geq q_i'(u_i)$ by concavity. Moreover, at least one of these two inequalities is strict unless $\rho_i = u_i$ and $q_i'(\rho_i) = q_p'(\rho_p)$, which implies that $\rho_p = u_p$ by our tie-breaking rule for p.

Thus, the only remaining case is that $\rho_p = u_p$, a degenerate case. This is actually an example of a Type A local minimum (i.e., all of the variables $x_1^*, \ldots, x_s^*$ are at boundary values) so we need only verify the conditions of Lemma 5.14. Use the definition of $\overline{F}$ appropriate for that lemma, i.e., $\overline{F} = \{1, \ldots, s\} - F$.

Define a subset of G, namely G_0, of the values of i that satisfy $\rho_i = u_i$ and $q_i'(u_i) = q_p'(u_p)$. Note that $p \in G_0$; any additional elements would imply further degeneracy in the problem. Also, note that $G_0 \subset \overline{F}$.

First, we claim that if $i \in G_0$ and $j \in \overline{F} - G_0$, then

$$
q_i'(u_i) > q_j'(u_j).
$$

To establish this, we need to take cases on j. If $j \notin G$ then $j \notin F \cup G$, hence this follows immediately from (5.16). If $j \in G$, then $\rho_j \leq u_j$ and $q_j'(\rho_j) \leq q_p'(\rho_p)$, and one of these inequalities is strict (because $j \notin G_0$). Thus, we have the following chain of inequalities, one of which must be strict:

$$
q_j'(u_j) \leq q_j'(\rho_j) \leq q_p'(\rho_p) = q_p'(u_p)
$$

which establishes the inequality.

This inequality shows that it is sufficient to prove (i), (ii) and (iii) of Lemma 5.14 substituting G_0 in place of $\overline{F}$ in the lemma.

Then we notice condition (i) follows immediately from (5.15). Next, we observe by (5.17) that $q_p'(\rho_p) \geq \psi'(\omega^*)$ (for some value of the right hand side); combining this with the fact that $q_i'(x_i^*) > q_p'(\rho_p)$ for $i \in F$ establishes condition (ii). Finally, condition (iii) is established for $i \in G_0$ by noting that if $i \in G_0$, then $\rho_i = u_i$, and therefore no smaller value of ρ_i satisfied (5.17). This means that there is an infinite sequence $\epsilon_1, \epsilon_2, \ldots$ of positive numbers converging to zero such that $q_i'(\rho_i - \epsilon_j) < \psi'(\omega^* + \epsilon_j)$, which by continuity establishes condition (iii). ∎

We now say a few words about implementing Algorithm LOCAL-MIN2. On each iteration several of the ρ_i must be computed. This is the most time-consuming step; each ρ_i requires $O(t)$ arithmetic operations using the searching scheme described earlier (recall that t is the number of variables merged to get ω). On each iteration there are $O(s)$ values of ρ_i that need be evaluated, and there are at most s iterations of the algorithm. Therefore, the time required is $O(s^2t)$. The initial preprocessing to form ψ required $O(t \log t)$ steps. Thus, the total requirement is at most $O(n^3)$, where $n = s + t$. This can be brought down to $O(n(\log n)^2)$ using binary search and a technique due to Chazelle and Guibas [1985] from the computational geometry literature, but a discussion of these techniques is outside the scope of this book.

In this section and the previous two sections, we did not address the question of membership in **P**. This is because there is no natural decision problem that can be associated with the problem of computing a local minimum. For example, we could not pose a decision problem concerning the objective function value at a local minimum, since each local minimum for a particular instance might have a distinct value.

Exercises

5.1 Let QPLOCN be the following decision problem. Given a quadratic programming problem $(H, \mathbf{c}, A, \mathbf{b})$ with integer data, and given a rational point $\mathbf{x}^*$, is $\mathbf{x}^*$ **not** a local minimum? Show that QPLOCN is **NP**-complete.

5.2 Consider the problem of minimizing $y - x^2$ subject to $x^2 + y^2 \leq 1$ illustrated in Fig. 5.1. Prove that $(0, -1)$ is not a local minimum, but there are no feasible descent directions from this point.

5.3 Prove that the conditions stated in Lemma 5.2 are necessary for local minimality. (In the text it is proved that they are sufficient.)

5.4 Give a 2-variable example of concave QKP in which Algorithm LOCAL-MIN1 will pick a local minimum whose objective function value is very far above the global minimum of the objective function.

5.5 Suppose that Conditions 1 and 2 introduced in Section 5.4 hold at an iteration of Algorithm SIMUL. Let $m \in G$ be the index such that $q'_m(r_m) = w$. Show that the assignment

$$x_i^* = \begin{cases} l_i & \text{for } i \in F, \\ u_i & \text{for } i \notin F \cup \{m\}, \\ r_m & \text{for } i = m \end{cases}$$

satisfies the conditions of Lemma 5.2.

5.6 (The following problem is open, even in the concave case.) Find a polynomial-time algorithm for the following problem. Given an instance of QKP and given a feasible point $\mathbf{x}^{(0)}$, find a local minimum $\mathbf{x}^*$ such that $f(\mathbf{x}^*) \leq f(\mathbf{x}^{(0)})$, where f denotes the objective function for the instance.

5.7 Find a polynomial time algorithm for the following problem. Given a diagonal (indefinite) matrix D and a vector $\mathbf{c}$, find a global minimum for the objective function $\frac{1}{2}\mathbf{x}^T D\mathbf{x} + \mathbf{c}^T\mathbf{x}$ subject to $\mathbf{x} \in \Delta^{n-1}$ (simplex constraints). [Hint: Use parameterization to eliminate all but one of the convex variables. Use the fact that the global minimum of a concave function on a simplex must occur at one of the n vertices.]

6

COMPLEXITY IN THE BLACK-BOX MODEL

So far we have focused mainly on quadratic objective functions and special cases. In this chapter we consider general objective functions. We assume the objective function is available as a black box that, given a vector $\mathbf{x}$ in the feasible domain, returns $f(\mathbf{x})$. In this model it is possible to prove lower bound complexity results based solely on how much information an algorithm can obtain about the objective function. Accordingly, the subject matter of this chapter might be called "information-based optimization."

We assume for the purpose of this chapter that our function f accepts as arguments real numbers and returns real numbers. Models of computation were discussed in Section 2.4. When researchers develop information-based algorithms, they think in terms of working with full real numbers. This is why we assume a real-number model of computation.

Our first result is mainly negative: We show that the worst case complexity for global minimization in the information model is exponential in the number of variables and the number of digits of accuracy. In the following section, we turn to local minimization; in this case the dependence on the number of digits remains exponential, but the running time is now polynomial in the number of variables.

In the third section we turn to convex optimization. Here the results are very positive; there is an efficient information-based algorithm for this problem.

We remark that the subject of information and complexity has been studied, for example, by Traub, Wasilkowski and Woźniakowski [1988]. Algorithms for unconstrained minimization in this model have been studied by Dennis and Schnabel [1983].

The algorithms proposed by Dennis and Schnabel are intended to compute some kind of local minimum rather than a global minimum. Accordingly, the results of Section 6.2 are of greater relevance. The kinds of algorithms they study include model trust region methods and line-search methods. Generally these algorithms evaluate derivatives of the objective function.

Global minimization in the function evaluation model has been studied by Törn and Žilinskas [1989]. Most of the algorithms proposed by Törn and Žilinskas appear to have similar properties from the point of view of worst-case complexity, so the main difference between them is their behavior in practice. Another reference on optimization algorithms in this model—particularly for the convex case—is Nemirovsky and Yudin [1983]. Results from their book are presented in Section 6.3.

Because the focus in this chapter is on the complexity of the objective function, we limit consideration to the simple case of box constraints. Let U^n denote the n dimensional unit cube, that is, the n-fold product of the interval $U = [0, 1]$.

This chapter is meant to address general objective functions; a different way to generalize the results of previous chapters is to consider the case of a general polynomial objective function and polynomial constraints. Recently, theoretical methods have been proposed (Canny [1988]; Renegar [1989]) for solving systems of polynomial constraints over the real numbers. This theory is able to address problems much more general than optimization of polynomial functions. The algorithms, however, have very large asymptotic running times. In general, problems involving polynomial constraints are **NP**-hard or worse—see Exercise 2.5 and Exercise 4.1.

6.1. Global minimization

Assume f is specified as a black box. We start by determining what sort of global minimization problem can be solved in the information model. Suppose $f : U^n \to \mathbb{R}$ is arbitrary and we want to find a point $\mathbf{x} \in U^n$ such that $f(\mathbf{x})$ is minimum. If f is not continuous, then such an $\mathbf{x}$ may not exist. For example, consider $f : [0, 1] \to \mathbb{R}$ given by

$$f(x) = \begin{cases} -x & \text{for } 0 \le x < 0.5, \text{ or} \\ 0 & \text{for } 0.5 \le x \le 1. \end{cases}$$

This function has no global minimum. Accordingly, we restrict attention to the case that $f(\mathbf{x})$ is continuous.

Even for continuous functions, there is no finite bound on the number of steps required to find the optimum in the worst case. To see this, consider the family of continuous functions $f_{a,\epsilon}$ from U to $\mathbb{R}$ parameterized by two real numbers $a, \epsilon \in U$. The function is defined as follows: $f_{a,\epsilon}(a - \epsilon) = 0$, $f_{a,\epsilon}(a) = -1$, $f_{a,\epsilon}(a + \epsilon) = 0$, $f_{a,\epsilon}$ is linearly interpolated in between, and is zero for points outside $(a - \epsilon, a + \epsilon)$. (See Fig. 6.1.)

The global minimum of $f_{a,\epsilon}$ is at a and has value -1. Now, consider an algorithm that is designed to find a global minimum of continuous functions $U \to \mathbb{R}$. We claim that for any number of evaluations k made by the algorithm, there is some function in the family such that the algorithm cannot determine its global minimum. To see this, we note that for any particular sequence of test points $x_1, \ldots, x_k$ lying in U, there is a choice of a, ϵ such that $f_{a,\epsilon}(x_i) = 0$ for all the test points. If this particular $f_{a,\epsilon}$ were the input

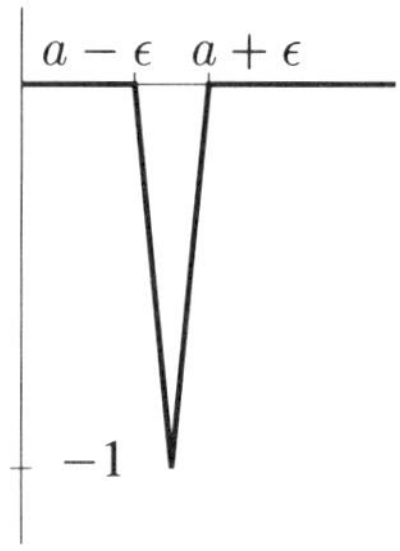

FIG. 6.1. A function whose minimum can be hidden.

to the optimization algorithm, then the algorithm could not distinguish $f_{a,\epsilon}$ from the function that is zero everywhere (and has a global minimum of 0).

At this point we should say more about the algorithms. We assume that on each iteration the algorithm can make one function evaluation at a test point. As mentioned in the introduction, the algorithm works with full real numbers. We allow the algorithm unlimited computational power (but it must be deterministic), and we allow the algorithm's test points to depend on previous test points—such an algorithm is sometimes called *adaptive*. Thus, the limit to the algorithm's power comes from the limited amount of information it can obtain about f.

We have just argued that there is no finite algorithm to find global minima even of continuous functions in the information model. The difficulty with the class of functions $f_{a,\epsilon}$ is their ability to "hide" a large jump in an arbitrarily small interval. We solve this problem by requiring f to have some regularity properties.

The simplest regularity property (and the property that most easily lends itself to analysis) is a Lipschitz condition. We assume that there is a constant L such that $|f(\mathbf{x}) - f(\mathbf{y})| \leq L\|\mathbf{x} - \mathbf{y}\|$ for all $\mathbf{x}, \mathbf{y}$ in U^n. We have to choose a norm for this property. In this section we use the ∞-norm because it lends itself most easily to analysis.

Note that $f_{a,\epsilon}$ has a Lipschitz bound—in fact, the bound is $L = 1/\epsilon$. The interesting case, therefore, is when the Lipschitz bound is fixed for the family of functions. Alternatively, we can attempt to get bounds on the running time that depend on L.

Even for continuous functions with a Lipschitz bound L there is no finite algorithm to find the global minimum exactly. In particular, we can make the following claim. After k function evaluations, there is always some closed interval $[a, b]$ of positive length such that the following condition holds for every $c \in [a, b]$. There is a Lipschitz continuous function f whose optimal value is c, and f is consistent with the first k function evaluations.

Thus, we reduce our goal to finding an approximate global minimum. There are two reasonable definitions for "approximate global minimum." The first definition is that $\mathbf{x}$ is an ϵ-approximate global minimum if there exists a global minimum $\mathbf{x}^*$ such that $\|\mathbf{x}-\mathbf{x}^*\|_\infty \leq \epsilon$. The second definition is that $\mathbf{x}$ is an ϵ-approximate global minimum if there exists a global minimum $\mathbf{x}^*$ such that $f(\mathbf{x}) - f(\mathbf{x}^*) \leq \epsilon$.

The first definition is stronger: The first definition plus Lipschitz continuity implies the second. The first definition, however, is unrealistic for actual computation. Consider, for example, the function $f : U \rightarrow \mathbb{R}$ given by $f(x) = 1 - 10^{-100}x$. This function is Lipschitz-continuous with $L = 10^{-100}$. Its unique global minimum is $x^* = 1$. In this example, the point $x = 0$ would be a 10^{-100} approximate global minimum in the second sense but not the first. In any realistic model of numerical computation, however, the two points would be indistinguishable and would both have objective function value of 1. Therefore, requiring that the optimization algorithm return a point close to $x^* = 1$ rather than 0 strains the plausibility of the real-number model of computation. Accordingly, we work only with the second definition. Let us state this formally.

DEFINITION. *Let $f : U^n \rightarrow \mathbb{R}$ be continuous. Then a point $\mathbf{x} \in U^n$ is said to be an* **ϵ-approximate global minimum** *if $f(\mathbf{x}) - f(\mathbf{x}^*) \leq \epsilon$, where $\mathbf{x}^*$ is a true global minimum.*

Now we can finally ask a question that has an answer: What is the worst-case complexity of finding an ϵ-approximate global minimum of a function $f : U^n \rightarrow \mathbb{R}$ that is Lipschitz continuous with constant L? Clearly the answer will depend on ϵ, L, n.

It turns out that this question is not too hard to answer— in some sense, our definitions have been set up so that it is easy to answer. The results we are about to prove are implicit in many places, such as Sikorski [1984] and the books mentioned at the end of this section.

We remark that none of our complexity bounds depend on L or ϵ individually; instead, they all depend on the ratio L/ϵ. This is expected because the problem of finding an ϵ-approximate global minimum for f is the same problem as finding a $c\epsilon$-approximate global minimum for cf (where $c > 0$). Therefore, we would expect the complexity to be unchanged if L and ϵ are scaled by the same amount.

We claim that the worst-case complexity grows like $(0.5L/\epsilon)^n$. We start by establishing this as an upper bound. One straightforward way to find an ϵ-approximation global minimum is to insert an n-dimensional mesh of points spaced $2\epsilon/L$ apart in each dimension. In particular, the point

$$((2i_1 + 1)\epsilon/L, \ldots, (2i_n + 1)\epsilon/L)$$

will be a point in the mesh, where $i_1, \ldots, i_n$ are integers between 0 and $0.5L/\epsilon$. There are at most $(0.5L/\epsilon + 1)^n$ points in this mesh. The algorithm is as follows: Evaluate f at each of these points, and then conclude that the ϵ-approximate global minimum occurs at the point with the smallest value of f.

To prove that this algorithm indeed returns an ϵ-approximate global minimum, consider a true global minimum $\mathbf{x}^*$. By choice of the grid, there is some grid point $\mathbf{x}$ such that $\|\mathbf{x}-\mathbf{x}^*\|_\infty \leq \epsilon/L$. Then the Lipschitz condition tells us that $|f(\mathbf{x}) - f(\mathbf{x}^*)| \leq \epsilon$, i.e., $f(\mathbf{x}) - f(\mathbf{x}^*) \leq \epsilon$. Since $\mathbf{x}$ is a grid point, the algorithm returns a point $\mathbf{x}_1$ whose objective function value is no greater than $f(\mathbf{x})$. This concludes the proof.

Now we turn to the lower bound, i.e., we attempt to prove that no algorithm can improve much on this naive grid algorithm in the worst case. Assume ϵ, L, n are chosen. Let ϵ' be slightly larger than ϵ. We focus on a particular class of functions $f_\mathbf{v} : U^n \to \mathbb{R}$ parameterized by a point $\mathbf{v} \in U^n$. We define

$$f_\mathbf{v}(\mathbf{x}) = \begin{cases} 0 & \text{if } \|\mathbf{v}-\mathbf{x}\|_\infty \geq \epsilon'/L, \text{ or} \\ L\|\mathbf{v}-\mathbf{x}\|_\infty - \epsilon' & \text{if } \|\mathbf{v}-\mathbf{x}\|_\infty \leq \epsilon'/L. \end{cases}$$

This function $f_\mathbf{v}$ has the following properties, which are easily verified. Its global minimum objective value is $-\epsilon'$ and is achieved at $\mathbf{x} = \mathbf{v}$. It is Lipschitz continuous with bound L.

Now, we claim that $(0.5L/\epsilon)^n$ function evaluations are required before an algorithm can distinguish all of the $f_\mathbf{v}$'s from the constant zero function. We allow any deterministic algorithm in the function-evaluation model. The algorithm is assumed to know ϵ, L, n. In particular, suppose that k function evaluations have been made, say $\mathbf{x}_1, \ldots, \mathbf{x}_k$, and suppose that $f(\mathbf{x}_i) = 0$ for all $i = 1, \ldots, k$. For which values of $\mathbf{v}$ is this sequence of function evaluations consistent with $f_\mathbf{v}$? Suppose $f_\mathbf{v}(\mathbf{x}_i) = 0$ for all i. This is true iff $\|\mathbf{x}_i - \mathbf{v}\|_\infty \geq \epsilon'/L$ for all $i = 1, \ldots, k$. Let $A_1, \ldots, A_k$ be sets defined as follows:

$$A_i = \{\mathbf{v} \in U^n : \|\mathbf{x}_i - \mathbf{v}\|_\infty < \epsilon'/L\}.$$

Thus, $f_\mathbf{v}$ is consistent with the data iff $\mathbf{v} \notin A_1 \cup \cdots \cup A_k$. The volume of each A_i is easily seen to be at most $(2\epsilon'/L)^n$. Therefore, the volume of $A_1 \cup \cdots \cup A_k$ is at most $k(2\epsilon'/L)^n$. We see that if $k < (0.5L/\epsilon')^n$, then $A_1 \cup \cdots \cup A_k$ has volume less than 1. Thus, if $k < (0.5L/\epsilon')^n$ then there is a point $\mathbf{v} \in U^n$ such that $f_\mathbf{v}$ is zero at all the test points so far. This means that the optimization algorithm cannot distinguish $f_\mathbf{v}$ from the function that is zero everywhere.

This means that if the algorithm reports that the approximate value of the global minimum is nonnegative when $k < (0.5L/\epsilon')^n$, then it would give the wrong answer for some $f_\mathbf{v}$ (because the global minimum is $-\epsilon'$ which is larger than the allowable ϵ tolerance). On the other hand, if the algorithm reports that the approximate global minimum is negative, then it would give the wrong answer if the function is the zero function.

Thus, at least $(0.5L/\epsilon')^n$ function evaluations are required. This holds for all $\epsilon' > \epsilon$, so it must hold for ϵ as well.

Therefore, we have shown that the worst-case complexity lies between $(0.5L/\epsilon)^n$ and $(0.5L/\epsilon + 1)^n$ for all ϵ, L, n. Let us state this as a theorem.

THEOREM 6.1. *The worst-case complexity of computing an ϵ-approximate global minimum for functions $f : U^n \to \mathbb{R}$ with Lipschitz constant L lies between $(0.5L/\epsilon)^n$ and $(0.5L/\epsilon + 1)^n$.*

It is interesting to compare this bound with the bounds for the Renegar-Shub algorithm of Chapter 3 and the Ye-Karmarkar algorithm of Chapter 4. In both cases the complexity grew as a polynomial in n and in $|\log \epsilon|$ (because the number of digits of accuracy grew proportionally to the number of iterations, and the number of digits of accuracy corresponds to $|\log \epsilon|$). For the present problem, any algorithm is exponential in both n and $|\log \epsilon|$. We return to this subject at the end of Section 6.3.

The other notable thing about these results is that allowing adaptive algorithms does not seem to help in terms of worst-case complexity. The algorithm proposed above is essentially the most simple-minded algorithm possible for the problem, and we have proved it is almost optimal in the worst case.

Results like this are also proved in Novak [1988], who establishes many other results about global optimization and about related problems. In this information-based model, other topics besides worst-case complexity may be studied. See Traub, Wasilkowski and Woźniakowski [1988] for discussions of randomized algorithms, average-case complexity and many other topics related to information-based complexity. Nemirovsky and Yudin [1983] consider the case that higher smoothness is assumed of the objective function.

6.2. Local minimization

In contrast to global minimization, the problem of finding an approximate local minimum, under suitable definitions, has complexity that is polynomial in n. We assume as in the last section that we are in a black-box model. The black-box model for local optimization is implicit in many of the algorithms for unconstrained local minimization, including the steepest descent method, the Powell-symmetric-Broyden method, the Broyden-Fletcher-Goldfarb-Shanno method and the line-search and trust-region modifications of these algorithms. See Dennis and Schnabel [1983] for a description of these algorithms. In this section we will discuss constrained optimization rather than unconstrained because the complexity issues appear to be simpler. If we focus on functions $f : U^n \to \mathbb{R}$, then there is a built in "scale" for the problem, and there is also knowledge that a local minimum must exist. Unconstrained functions $f : \mathbb{R}^n \to \mathbb{R}$ do not necessarily have local minima.

As noted previously, it is not possible to find an exact local minimum, so we must define what is meant by an approximate local minimum. We propose the following definition. This definition, as well as the rest of the material in this section, first appeared in Vavasis [1990c].

DEFINITION. *A point* $\mathbf{x}^* \in U^n$ *is said to be an* **ϵ-approximate local minimum** *of a continuous function* $f : U^n \to \mathbb{R}$ *if there exists an open set* N *containing* $\mathbf{x}^*$ *such that*

$$f(\mathbf{x}^*) \leq f(\mathbf{x}) + \epsilon\|\mathbf{x} - \mathbf{x}^*\|$$

for all $\mathbf{x} \in N \cap U^n$.

Below we give an alternate characterization of this definition. First, we explain this definition and also point out its shortcomings. The motivation of this definition is that while $\mathbf{x}^*$ may not have the smallest function value in the neighborhood N, the value of f decreases slowly (at a rate no faster than ϵ) as one moves away from $\mathbf{x}^*$.

The most obvious shortcoming of this definition is that an interior local maximum or interior saddle point would also qualify as an ϵ-approximate local minimum under this definition. We do not feel that this property is a severe flaw in the definition, however. For example, examining the local minimization algorithms of Dennis and Schnabel, one sees that it is possible for these algorithms to converge to local maxima. Indeed, distinguishing local minima from other kinds of stationary points is a computationally difficult problem; a special case of this problem was shown to be **NP**-hard in Section 5.1.

We observe that a norm must be selected in the above definition. For this section we will assume that the 1-norm is used in that definition. The norms in this section have been selected to make the analysis simple.

We now give an alternative characterization of an approximate local minimum. We will say that $\mathbf{x}^* = (x_1^*, \ldots, x_n^*)^T$ is an *ϵ-KKT point* of $f : U^n \to \mathbb{R}$ if

1. For all i such that $x_i^* > 0$, $\partial f/\partial x_i(\mathbf{x}^*) \leq \epsilon$, and

2. For all i such that $x_i^* < 1$, $\partial f/\partial x_i(\mathbf{x}^*) \geq -\epsilon$.

(Note that if $\epsilon = 0$ these conditions are the KKT conditions.) If $\mathbf{x}^*$ is interior, these conditions are equivalent to the requirement that $\|\nabla f(\mathbf{x}^*)\|_\infty \leq \epsilon$.

Assuming f is differentiable, we claim that if a point $\mathbf{x}^*$ is an ϵ-approximate local minimum, then it is an ϵ-KKT point. We verify statement 1 above for a particular index i (statement 2 is similar). Assuming $x_i^* > 0$, the point $\mathbf{x}^* - t\mathbf{e}_i$ is feasible for small enough $t > 0$, where $\mathbf{e}_i$ is the ith column of the identity matrix. For small enough t, this point is contained in $N \cap I$, hence $f(\mathbf{x}^*) - f(\mathbf{x}^* - t\mathbf{e}_i) \leq t\epsilon$ by definition of approximate local minimum. Since this holds for all t small enough, by definition of the partial derivative this implies $\partial f/\partial x_i(\mathbf{x}^*) \leq \epsilon$.

Conversely, suppose $\mathbf{x}^*$ is an ϵ-KKT point. Then we claim it is an ϵ'-approximate local minimum for any $\epsilon' > \epsilon$. To see this, recall that the definition of a derivative is that for all $\mathbf{d}$,

$$f(\mathbf{x}^* + \mathbf{d}) = f(\mathbf{x}^*) + \nabla f(\mathbf{x}^*)^T\mathbf{d} + o(\|\mathbf{d}\|).$$

Suppose that $x_i^* > 0$ for some i; then we know

$$f(\mathbf{x}^* - t\mathbf{e}_i) = f(\mathbf{x}^*) - t\frac{\partial f}{\partial x_i}(\mathbf{x}^*) + o(t)$$

so

$$f(\mathbf{x}^* - t\mathbf{e}_i) \geq f(\mathbf{x}^*) - t\epsilon + o(t)$$

so

$$f(\mathbf{x}^* - t\mathbf{e}_i) \geq f(\mathbf{x}^*) - t\epsilon'$$

for all t small enough. This inequality still holds with $\mathbf{x}$ substituted for $\mathbf{x}^*$, for all $\mathbf{x}$ in a neighborhood of $\mathbf{x}^*$, since we are assuming that ∇f is continuous. Then we see that we can get a lower bound on $f(\mathbf{x}^* + \mathbf{d})$ for an arbitrary $\mathbf{d}$ that is small enough by expressing $\mathbf{d}$ as a sum of small steps of the form $t_i\mathbf{e}_i$.

We next ask the question: Given a continuously differentiable function $f : U^n \to \mathbb{R}$ and given a number $\epsilon > 0$, what is the complexity of finding an ϵ-approximate local minimum? It turns out that the there is no finite upper bound on the number of steps. In particular, for any finite sequence of test points $x_1, \ldots, x_k$, there exists a continuously differentiable function $f : [0, 1] \to \mathbb{R}$ such that $f(x_i) = 0$ and $f'(x_i) = 1$ at all test points (except if $x_i = 0$ then $f'(0) = -1$). Moreover, $-1 \leq f'(x) \leq 1$ for all $x \in [0, 1]$ which implies that f has a Lipschitz bound of 1. Figure 6.2 illustrates an example of a sequence of test points and the bad-case function for these points. To construct this function, put the x_i's into increasing order, and then let f be a correctly chosen cubic Hermite function based on the x_i's.

An algorithm trying to find approximate local minima for this family of functions will always completely fail (i.e., it will discover that $f(x) = 0$ and $f'(x) = 1$ at all of its test points) for at least one function in the family after any finite number of steps.

The problem with this family of functions is that the first derivatives can vary too much over short intervals, so that the algorithm is prevented from obtaining useful information on the first derivative of the function.

Accordingly, we place an additional restriction on the function. In particular, we require that the first derivative satisfy a Lipschitz condition, that is, there exists a constant M such that

$$\|\nabla f(\mathbf{x}) - \nabla f(\mathbf{y})\|_\infty \leq M\|\mathbf{x} - \mathbf{y}\|_\infty$$

for all $\mathbf{x}, \mathbf{y} \in U^n$.

We now ask the question, what is the complexity of finding an ϵ-approximate local minimum for a function in this class? Clearly the answer depends on ϵ, M, and n. We give an algorithm for this problem, which yields an upper bound on the complexity. We call this algorithm INFO-LOCAL.

We assume that ϵ, M and n are given. We assume also that M/ϵ is an integer. We are given a starting point $\mathbf{x}^{(0)} \in U^n$, which is assumed to have

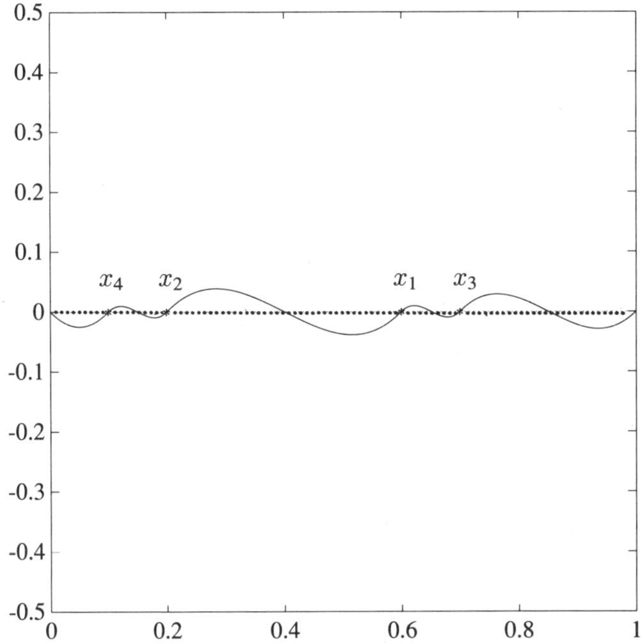

FIG. 6.2. An impossible case for local minimization.

each coordinate equal to an integer multiple of ϵ/M. If no starting point is given, the origin can be used.

Given a function $f(\mathbf{x})$, we define vector-valued function $\mathbf{g}(\mathbf{x})$ as follows. The ith entry of $\mathbf{g}(\mathbf{x})$ is defined by

$$g_i(\mathbf{x}) = \begin{cases} \min\left(0, \dfrac{\partial f}{\partial x_i}(\mathbf{x})\right) & \text{if } x_i = 0, \\ \dfrac{\partial f}{\partial x_i}(\mathbf{x}) & \text{if } 0 < x_i < 1, \text{ or} \\ \max\left(0, \dfrac{\partial f}{\partial x_i}(\mathbf{x})\right) & \text{if } x_i = 1. \end{cases}$$

Notice that if $\mathbf{x}$ is interior to U^n, then $\mathbf{g}(\mathbf{x}) = \nabla f(\mathbf{x})$. This function $\mathbf{g}(\mathbf{x})$ could be called the "projected gradient," although this terminology is not standard.

Algorithm INFO-LOCAL begins by testing whether $\|\mathbf{g}(\mathbf{x}^{(0)})\|_\infty > M$. If this inequality holds, then for some i we know $|\partial f/\partial x_i(\mathbf{x}^{(0)})| > M$. Take the case that $\partial f/\partial x_i(\mathbf{x}^{(0)}) > M$ (the negative case is similar). We claim that $\partial f/\partial x_i(\mathbf{x}) > 0$ for all $\mathbf{x} \in U^n$. This follows from the Lipschitz bound on ∇f.

This means that any local minimum of f must occur on the face $T =$

$\{\mathbf{x} \in U^n : x_i = 0\}$ of U^n. Moreover, if $f_0 : T \to \mathbb{R}$ denotes the restriction of f to T, then it suffices to find an ϵ-approximate local minimum of f_0. Therefore, we can project $\mathbf{x}^{(0)}$ onto T and work on the restricted problem. The restriction operation has the effect of deleting the ith entry from the vector $\mathbf{g}(\mathbf{x})$.

Accordingly, we can continue to reduce the dimensionality of the problem coordinate by coordinate. Therefore, without loss of generality, we can assume our starting point satisfies $\|\mathbf{g}(\mathbf{x}^{(0)})\|_\infty \le M$.

Let $\mathbf{x}^*$ be a global minimum of f. We can use the upper bound on $\mathbf{g}$ to derive an upper bound on the difference $f(\mathbf{x}^{(0)}) - f(\mathbf{x}^*)$. Let $s = \|\mathbf{x}^* - \mathbf{x}^{(0)}\|_1$. Then we can construct a path made up of segments parallel to the coordinate axes from $\mathbf{x}^{(0)}$ to $\mathbf{x}^*$; the length of this path will be exactly s. Assume the path is made up of n segments $P_1, \ldots, P_n$ such that P_i is parallel to $\mathbf{e}_i$.

Then we know that

$$f(\mathbf{x}^*) - f(\mathbf{x}^{(0)}) = \sum_{i=1}^{n} \int_{P_i} \frac{\partial f}{\partial x_i}. \tag{6.1}$$

We now derive a lower bound on each integral in (6.1). There are two cases. In the first case, $g_i(\mathbf{x}^{(0)}) = \partial f / \partial x_i(\mathbf{x}^{(0)})$. In this case, we can apply the Lipschitz bound directly. We know that $|g_i(\mathbf{x}^{(0)})| = |\partial f / \partial x_i(\mathbf{x}^{(0)})| \le M$. Since the distance from $\mathbf{x}^{(0)}$ to any point of P_i is at most 1 in the ∞-norm, we know that the magnitude of $\partial f / \partial x_i$ along P_i is at most $2M$. Therefore, the ith integral on the right-hand side of (6.1) has magnitude at most $2M$.

In the second case, $g_i(\mathbf{x}^{(0)}) \neq \partial f / \partial x_i(\mathbf{x}^{(0)})$. Examining the definition of $\mathbf{g}$, we see that there are two possible subcases: either $x_i^{(0)} = 0$ and $\partial f / \partial x_i(\mathbf{x}^{(0)}) > 0$, or $x_i^{(0)} = 1$ and $\partial f / \partial x_i(\mathbf{x}^{(0)}) < 0$. We treat the first subcase since the second subcase is analogous. Since $x_i^{(0)} = 0$ and $\partial f / \partial x_i(\mathbf{x}^{(0)}) > 0$, we know that $\partial f / \partial x_i$ cannot drop below $-M$ at any point on P_i. Moreover, we know that P_i is oriented in the positive direction with respect to $\mathbf{e}_i$, because $x_i^{(0)} = 0$. Therefore, the ith integral in the above summation is at least $-M$ (this argument does not give an upper bound, but only a lower bound is needed).

We conclude that all the integrals in (6.1) are at least $-2M$, and hence

$$f(\mathbf{x}^*) - f(\mathbf{x}^{(0)}) \ge -2Mn,$$

i.e.,

$$f(\mathbf{x}^{(0)}) - f(\mathbf{x}^*) \le 2Mn.$$

This gives an upper bound on how much the objective function can decrease.

We now continue the description of INFO-LOCAL, under the assumption that $\|\mathbf{g}(\mathbf{x}^{(0)})\|_\infty \le M$. The algorithm operates on an imaginary grid of nodes spaced ϵ/M apart in each dimension of U^n and aligned with the coordinate axes. According to our earlier assumptions, there are an integer number of mesh cells in every dimension, and the initial point $\mathbf{x}^{(0)}$ is one of the mesh points.

We use the following iteration. Assume the current iterate is $\mathbf{x}^{(k)}$. We compute $\mathbf{g}(\mathbf{x}^{(k)})$. If $\|\mathbf{g}(\mathbf{x}^{(k)})\|_\infty < \epsilon$, then we halt. The justification for halting is as follows. If $\|\mathbf{g}(\mathbf{x}^{(k)})\|_\infty = \epsilon'$ and $\epsilon' < \epsilon$, then it is easy to verify by definition of $\mathbf{g}$ that $\mathbf{x}^{(k)}$ is an ϵ'-KKT point, and is therefore an ϵ-approximate local minimum.

Otherwise, suppose $\|\mathbf{g}(\mathbf{x}^{(k)})\|_\infty \geq \epsilon$. Then we identify a component, say $g_i(\mathbf{x}^{(k)})$, whose absolute value is at least ϵ. Say, for example, that $g_i(\mathbf{x}^{(k)}) \geq \epsilon$. This means by definition that $\partial f/\partial x_i(\mathbf{x}^{(k)}) \geq \epsilon$ and that $x_i^{(k)} > 0$. Then we set $\mathbf{x}^{(k+1)} = \mathbf{x}^{(k)} - (\epsilon/M)\mathbf{e}_i$. If $g_i(\mathbf{x}^{(k)})$ is negative, then we would instead add $(\epsilon/M)\mathbf{e}_i$. Notice that, under this definition, $\mathbf{x}^{(k+1)}$ will be a mesh point lying in U^n.

With this formula for $\mathbf{x}^{(k+1)}$, we claim that $f(\mathbf{x}^{(k+1)}) \leq f(\mathbf{x}^{(k)}) - 0.5\epsilon^2/M$. To see this, observe that in the case that $\partial f/\partial x_i(\mathbf{x}^{(k)})$ is positive,

$$\begin{aligned} f(\mathbf{x}^{(k)}) - f(\mathbf{x}^{(k+1)}) &= \int_{-\epsilon/M}^{0} \frac{\partial f}{\partial x_i}(\mathbf{x}^{(k)} + t\mathbf{e}_i)\,dt \\ &\geq \int_{-\epsilon/M}^{0} (\epsilon + Mt)\,dt \\ &\geq 0.5\epsilon^2/M. \end{aligned}$$

To derive the second line, we used the fact that $\partial f/\partial x_i(\mathbf{x}^{(k)}) \geq \epsilon$ as well as the Lipschitz bound on $\partial f/\partial x_i$.

We conclude that the objective function decreases each iteration by at least $0.5\epsilon^2/M$. As noted earlier, the most that the objective function can decrease is $2Mn$. Therefore, the maximum number of iterations is $4n(M/\epsilon)^2$. Let us state this as a theorem.

THEOREM 6.2. *Let $f : U^n \to \mathbb{R}$ be a C^1 function whose gradient satisfies a Lipschitz condition with bound M. Then an ϵ-approximate local minimum can be found with at most $4n(M/\epsilon)^2$ function and gradient evaluations.*

We remark that if gradient values are not available, the first part of the algorithm (the restrictions to subproblems) can be carried out by estimating the gradient via finite differences. A bound can be derived on the accuracy of finite difference approximations to the gradient using the hypothesis that the gradient is Lipschitz-bounded.

If gradient values are not available, then the main local-search step of the algorithm can be replaced with a comparison of the objective value at $\mathbf{x}^{(k)}$ to the objective values at the neighboring grid points. This requires $2n$ function evaluations per local search step.

In the last section we were able to give upper and lower complexity bounds for global minimization that were nearly matching. The bounds known for local minimization are not as close. In Vavasis [1990c] a lower bound of $\Omega((M/\epsilon)^{1/2})$ is proved for the case $n = 2$. Also an upper bound of the form $O(M/\epsilon)$ is proved when n is very small with respect to M/ϵ (this improves on the $(M/\epsilon)^2$ factor given by the previous theorem). We conclude that logarithmic dependence on M/ϵ is out of the question if $n \geq 2$ (for the case $n = 1$, see Exercise 6.4). An exact characterization of how the worst-case complexity varies with n and M/ϵ is not known.

6.3. Convex minimization and the ellipsoid algorithm

In the previous two sections, we observed polynomial dependence on L/ϵ, where L was a Lipschitz constant. In this section we introduce an efficient algorithm for minimization of a convex function f. This algorithm has logarithmic dependence on L/ϵ. We return to the definitions of Section 6.1, i.e., we are searching for an approximate global minimum as defined in that section. For this section we assume f is convex and has Lipschitz bound L.

The algorithm we will discuss is called the *ellipsoid* algorithm. We will present the algorithm specialized to the case of box constraints, but it applies to essentially arbitrary convex optimization problems. The material in this section is based on a much more general treatment by Nemirovsky and Yudin [1983], who refer to the algorithm as the "modified method of centers of gravity" (MMCG). It was first proposed in a paper (Iudin and Nemirovskii [1976]) by the same authors, and turned out to be a variation of an earlier method proposed by Shor [1970].

Assume that we have a differentiable convex function $f : U^n \to \mathbb{R}$. The assumption of differentiability can be dropped; in place of ∇f we could use the *subdifferential* of f. (Recall that the subdifferential for a convex function of one variable was introduced in Chapter 5.) We do not pursue this generalization here.

We assume that the function f and its gradient ∇f can be evaluated at arbitrary points in U^n. In the last section, we saw that the gradient of the objective function is useful because it gives a local direction of descent. In the convex case, the gradient $\nabla f(\mathbf{x})$ gives *global* information about the function. In particular, we have the following theorem.

THEOREM 6.3. *Let D be a nonempty convex subset of $\mathbb{R}^n$, and let f be a convex function on D and $\mathbf{x}$ a point of D. If $\mathbf{y} \in D$ is a point such that $f(\mathbf{y}) \leq f(\mathbf{x})$, then*

$$\nabla f(\mathbf{x})^T(\mathbf{y} - \mathbf{x}) \leq 0.$$

In particular, if a minimizer $\mathbf{x}^*$ exists, then $\nabla f(\mathbf{x})^T(\mathbf{x}^* - \mathbf{x}) \leq 0$.

PROOF. Assume $\mathbf{y} \neq \mathbf{x}$ (otherwise the theorem is trivial). For $\lambda \in [0, 1]$, consider

$$\mathbf{x}(\lambda) = (1 - \lambda)\mathbf{x} + \lambda\mathbf{y}.$$

By convexity, $\mathbf{x}(\lambda) \in D$ and

$$\begin{aligned} f(\mathbf{x}(\lambda)) &\leq (1-\lambda)f(\mathbf{x}) + \lambda f(\mathbf{y}) \\ &\leq f(\mathbf{x}). \end{aligned}$$

On the other hand, by definition of derivative,

$$\begin{aligned} f(\mathbf{x}(\lambda)) - f(\mathbf{x}) &= \nabla f(\mathbf{x})^T(\mathbf{x}(\lambda) - \mathbf{x}) + o(\lambda) \\ &= \lambda \nabla f(\mathbf{x})^T(\mathbf{y} - \mathbf{x}) + o(\lambda). \end{aligned}$$

Since the left-hand side is nonpositive for all $\lambda > 0$, then $\nabla f(\mathbf{x})^T(\mathbf{y}-\mathbf{x}) \leq 0$. ∎

Thus, from evaluating $\nabla f(\mathbf{x})$ we can construct a halfspace

$$\{\mathbf{y} \in \mathbb{R}^n : \nabla f(\mathbf{x})^T(\mathbf{y} - \mathbf{x}) \leq 0\}$$

that is guaranteed to contain the minimizer.

We now begin a description of the ellipsoid algorithm. We start with an ellipsoid E_1 containing the entire feasible region. Recall that an *ellipsoid* is defined to be a set of the form:

$$\{\mathbf{x} \in \mathbb{R}^n : (\mathbf{x} - \mathbf{c})^T B^{-1}(\mathbf{x} - \mathbf{c}) \leq 1\}$$

where B is an $n \times n$ symmetric positive definite matrix. Here, $\mathbf{c}$ is a vector called the *center* of the ellipsoid.

We can take E_1 to be the ellipsoid with center $\mathbf{c}^{(1)} = (1/2, \ldots, 1/2)$ and $B_1 = nI/4$, where I denotes the $n \times n$ identity matrix. It is easy to verify that every point $\mathbf{x}$ in U^n satisfies $4(\mathbf{x} - \mathbf{c}^{(1)})^T(\mathbf{x} - \mathbf{c}^{(1)})/n \leq 1$.

The invariant properties of the algorithm are that the sequence of ellipsoids $E_1, E_2, \ldots$ have decreasing volume, and that each one contains the minimizer.

The algorithm also maintains a set F of the objective function values evaluated at $\mathbf{c}^{(i)}$ (the center of the ith ellipsoid) for each i such that the center is feasible.

At the ith step of the ellipsoid algorithm, we check whether $\mathbf{c}^{(i)}$ is feasible. The first case is that $\mathbf{c}^{(i)}$ is infeasible. In this case, we find a constraint violated by the point. Since all the constraints of our problem take the form $x_j \geq 0$ and $x_j \leq 1$, then for some j either $c_j^{(i)} < 0$ or $c_j^{(i)} > 1$. We will treat the latter case; the former is handled in a similar manner. If $c_j^{(i)} > 1$, then we consider the halfspace

$$H = \{\mathbf{x} \in \mathbb{R}^n : x_j \leq c_j^{(i)}\}.$$

Clearly the minimizer is contained in $E_i \cap H$. Thus, we want ellipsoid E_{i+1} to contain $E_i \cap H$.

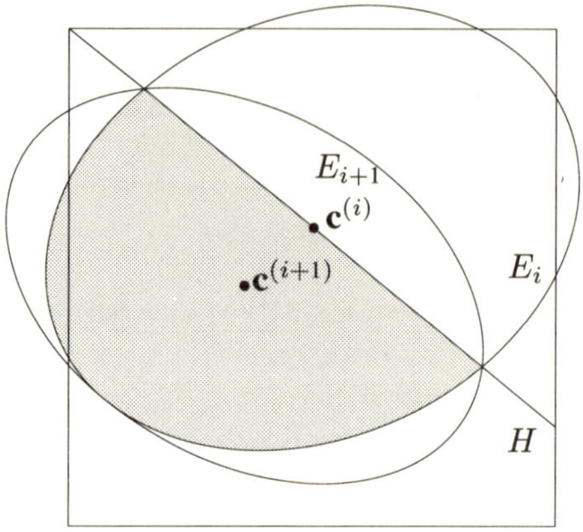

FIG. 6.3. One step of the ellipsoid algorithm.

The second case is that $\mathbf{c}^{(i)}$ is feasible. In this case, we compute $f(\mathbf{c}^{(i)})$ and append this number to set F. Then we compute $\nabla f(\mathbf{c}^{(i)})$. If this is the zero vector, then we have found the exact minimizer and the algorithm halts. Otherwise, we know from the theorem that the minimizer lies in the halfspace

$$H = \{\mathbf{x} \in \mathbb{R}^n : \nabla f(\mathbf{c}^{(i)})^T(\mathbf{x} - \mathbf{c}^{(i)}) \leq 0\}.$$

Thus, in either case, we end up with a halfspace H such that $E_i \cap H$ contains the minimizer. Moreover, in either case, the hyperplane that defines H passes through $\mathbf{c}^{(i)}$. (In the first case, the hyperplane is $\{x_j = c_j^{(i)}\}$. In the second case, the hyperplane is $\{\nabla f(\mathbf{c}^{(i)})^T(\mathbf{x} - \mathbf{c}^{(i)}) = 0\}$.)

We now explain the procedure to determine E_{i+1}. The formulas in this and the next few paragraphs follow the notation of Bland, Goldfarb and Todd [1981]. In both cases described above let halfspace H be specified by the inequality $\mathbf{a}^T(\mathbf{x} - \mathbf{c}^{(i)}) \leq 0$, where $\mathbf{a}$ is a nonzero vector. Then the formula for E_{i+1} is

$$E_{i+1} = \{\mathbf{x} : (\mathbf{x} - \mathbf{c}^{(i+1)})^T B_{i+1}^{-1}(\mathbf{x} - \mathbf{c}^{(i+1)}) \leq 1\}$$

where

$$\mathbf{c}^{(i+1)} = \mathbf{c}^{(i)} - \frac{\tau B_i \mathbf{a}}{\sqrt{\mathbf{a}^T B_i \mathbf{a}}}$$

and

$$B_{i+1} = \delta\left(B_i - \frac{\sigma B_i \mathbf{a}\mathbf{a}^T B_i}{\mathbf{a}^T B_i \mathbf{a}}\right).$$

In these formulas, $\tau = 1/(n+1)$, $\sigma = 2/(n+1)$, and $\delta = n^2/(n^2-1)$. A two-dimensional illustration of E_i, E_{i+1}, and H are in Fig. 6.3. In this figure, $E_i \cap H$ is shaded.

Before starting on the main claim about the ellipsoid algorithm, we first establish the following straightforward result from linear algebra.

LEMMA 6.1. *Let* $\mathbf{x}, \mathbf{y}$ *be vectors in* $\mathbb{R}^n$, *and let* B *be an* $n \times n$ *symmetric positive definite matrix. Then*

$$(\mathbf{x}^T B\mathbf{x})(\mathbf{y}^T B\mathbf{y}) \geq (\mathbf{x}^T B\mathbf{y})^2. \tag{6.2}$$

PROOF. Let the Choleski factorization of B be $B = LL^T$. Let $\mathbf{u} = L^T\mathbf{x}$ and $\mathbf{v} = L^T\mathbf{y}$. Then the left-hand side of (6.2) is $(\mathbf{u}^T\mathbf{u})(\mathbf{v}^T\mathbf{v})$. The right-hand side is $(\mathbf{u}^T\mathbf{v})^2$. Thus, the (6.2) follows from the Schwarz inequality. ∎

We have the following claim about these formulas.

LEMMA 6.2. *This formula for* E_i *has the following properties:*

1. B_{i+1} *is symmetric positive definite, and*
2. $E_i \cap H \subset E_{i+1}$.

PROOF. First, we argue that B_{i+1} is symmetric and positive definite. The symmetry is clear from the update formula, since both terms are symmetric. For positive definiteness, we compute:

$$\begin{aligned}
\mathbf{x}^T B_{i+1}\mathbf{x} &= \delta\left(\mathbf{x}^T B_i\mathbf{x} - \frac{\sigma\mathbf{x}^T B_i\mathbf{a}\mathbf{a}^T B_i\mathbf{x}}{\mathbf{a}^T B_i\mathbf{a}}\right) \\
&\geq \delta\left(\mathbf{x}^T B_i\mathbf{x} - \frac{\sigma(\mathbf{x}^T B_i\mathbf{x})(\mathbf{a}^T B_i\mathbf{a})}{\mathbf{a}^T B_i\mathbf{a}}\right) \\
&\geq \delta\left(\mathbf{x}^T B_i\mathbf{x} - \sigma\mathbf{x}^T B_i\mathbf{x}\right).
\end{aligned}$$

Since B_i is positive definite and $\sigma < 1$, the claim that B_{i+1} is positive definite follows. Note that we used (6.2) to derive the second line from the first.

Next, we have to prove the claim that $E_i \cap H \subset E_{i+1}$. First, we recall the *Sherman-Morrison* formula. If B is a nonsingular $n \times n$ matrix and $\mathbf{w}, \mathbf{y} \in \mathbb{R}^n$, then $B - \mathbf{w}\mathbf{y}^T$ is invertible provided $\mathbf{y}^T B^{-1}\mathbf{w} \neq 1$. The inverse is given by:

$$(B - \mathbf{w}\mathbf{y}^T)^{-1} = B^{-1} + \alpha^{-1}B^{-1}\mathbf{w}\mathbf{y}^T B^{-1}$$

where $\alpha = 1 - \mathbf{y}^T B^{-1}\mathbf{w}$. We use this formula to obtain a formula for B_{i+1}^{-1}:

$$\begin{aligned}
B_{i+1}^{-1} &= \delta^{-1}\left[B_i^{-1} + \sigma\alpha^{-1}\frac{B_i^{-1}B_i\mathbf{a}\mathbf{a}^T B_i B_i^{-1}}{\mathbf{a}^T B_i\mathbf{a}}\right] \\
&= \delta^{-1}\left[B_i^{-1} + \sigma\alpha^{-1}\frac{\mathbf{a}\mathbf{a}^T}{\mathbf{a}^T B_i\mathbf{a}}\right]
\end{aligned}$$

where

$$\begin{aligned}
\alpha &= 1 - \sigma\frac{\mathbf{a}^T B_i B_i^{-1} B_i\mathbf{a}}{\mathbf{a}^T B_i\mathbf{a}} \\
&= 1 - \sigma.
\end{aligned}$$

Let us consider a point $\mathbf{x} \in E_i \cap H$:

$$(\mathbf{x} - \mathbf{c}^{(i+1)})B_{i+1}^{-1}(\mathbf{x} - \mathbf{c}^{(i+1)}) = \delta^{-1}(\mathbf{u} + \mathbf{v})^T(B_i^{-1} + C)(\mathbf{u} + \mathbf{v})$$

where $\mathbf{u} = \mathbf{x} - \mathbf{c}^{(i)}$, $\mathbf{v} = \tau B_i\mathbf{a}/\sqrt{\mathbf{a}^T B_i \mathbf{a}}$, and $C = (\sigma \mathbf{a}\mathbf{a}^T)/((1-\sigma)\mathbf{a}^T B_i \mathbf{a})$. The goal for the upcoming calculations is to show that this expression is at most 1, proving that $\mathbf{x} \in E_{i+1}$. We break this up as

$$\delta^{-1}(t_1 + t_2 + t_3 + t_4 + t_5 + t_6)$$

where $t_1, \ldots, t_6$ are analyzed in the upcoming paragraphs.

First,

$$\begin{aligned} t_1 &= \mathbf{u}^T B_i^{-1}\mathbf{u} \\ &= (\mathbf{x} - \mathbf{c}^{(i)})^T B_i^{-1}(\mathbf{x} - \mathbf{c}^{(i)}) \\ &\leq 1 \end{aligned}$$

since $\mathbf{x} \in E_i$.

Second, we evaluate:

$$\begin{aligned} t_2 &= 2\mathbf{u}^T B_i^{-1}\mathbf{v} \\ &= 2\tau(\mathbf{x} - \mathbf{c}^{(i)})^T B_i^{-1}\frac{B_i\mathbf{a}}{\sqrt{\mathbf{a}^T B_i\mathbf{a}}} \\ &= \frac{2\tau(\mathbf{x} - \mathbf{c}^{(i)})^T\mathbf{a}}{\sqrt{\mathbf{a}^T B_i\mathbf{a}}}. \end{aligned}$$

Let μ denote $(\mathbf{x}-\mathbf{c}^{(i)})^T\mathbf{a}/\sqrt{\mathbf{a}^T B_i\mathbf{a}}$. Then $t_2 = 2\tau\mu$. We claim that $\mu \in [-1, 0]$. The inequality $\mu \leq 0$ follows from the definition of H and the fact that $\mathbf{x} \in H$. The inequality $\mu \geq -1$ is derived as follows. Let LL^T be the Choleski factorization of B_i. We have

$$\begin{aligned} \left|(\mathbf{x} - \mathbf{c}^{(i)})^T\mathbf{a}\right| &= \left|(\mathbf{x} - \mathbf{c}^{(i)})^T L^{-T}L^T\mathbf{a}\right| \\ &\leq \sqrt{(\mathbf{x} - \mathbf{c}^{(i)})^T L^{-T}L^{-1}(\mathbf{x} - \mathbf{c}^{(i)})} \cdot \sqrt{\mathbf{a}^T LL^T\mathbf{a}} \\ &\leq \sqrt{(\mathbf{x} - \mathbf{c}^{(i)})^T B_i^{-1}(\mathbf{x} - \mathbf{c}^{(i)})} \cdot \sqrt{\mathbf{a}^T B_i\mathbf{a}} \\ &\leq \sqrt{\mathbf{a}^T B_i\mathbf{a}}. \end{aligned}$$

This proves that $|\mu| \leq 1$.

Third, we have:

$$\begin{aligned} t_3 &= \mathbf{v}^T B_i^{-1}\mathbf{v} \\ &= \tau^2\frac{\mathbf{a}^T B_i}{\sqrt{\mathbf{a}^T B_i\mathbf{a}}}B_i^{-1}\frac{B_i\mathbf{a}}{\sqrt{\mathbf{a}^T B_i\mathbf{a}}} \\ &= \tau^2. \end{aligned}$$

Fourth, we evaluate:

$$\begin{aligned} t_4 &= \mathbf{u}^T C \mathbf{u} \\ &= (\mathbf{x} - \mathbf{c}^{(i)})^T \frac{\sigma \mathbf{a}\mathbf{a}^T}{(1-\sigma)\mathbf{a}^T B_i \mathbf{a}} (\mathbf{x} - \mathbf{c}^{(i)}) \\ &= \frac{\sigma \mu^2}{1-\sigma} \end{aligned}$$

by definition of μ.

Fifth,

$$\begin{aligned} t_5 &= 2\mathbf{u}^T C \mathbf{v} \\ &= 2\tau(\mathbf{x} - \mathbf{c}^{(i)})^T \left(\frac{\sigma \mathbf{a}\mathbf{a}^T}{(1-\sigma)\mathbf{a}^T B_i \mathbf{a}} \right) \left(\frac{B_i \mathbf{a}}{\sqrt{\mathbf{a}^T B_i \mathbf{a}}} \right) \\ &= \frac{2\tau\sigma\mu}{1-\sigma} \end{aligned}$$

by cancelling the $\mathbf{a}^T B_i \mathbf{a}$ in the numerator and denominator.

Sixth,

$$\begin{aligned} t_6 &= \mathbf{v}^T C \mathbf{v} \\ &= \tau^2 \left(\frac{\mathbf{a}^T B_i}{\sqrt{\mathbf{a}^T B_i \mathbf{a}}} \right) \left(\frac{\sigma \mathbf{a}\mathbf{a}^T}{(1-\sigma)\mathbf{a}^T B_i \mathbf{a}} \right) \left(\frac{B_i \mathbf{a}}{\sqrt{\mathbf{a}^T B_i \mathbf{a}}} \right) \\ &= \frac{\sigma\tau^2}{1-\sigma}. \end{aligned}$$

We want to sum up all the terms. Let us first add $t_2 + t_4 + t_5$. Notice that

$$\begin{aligned} t_2 + t_4 + t_5 &= 2\tau\mu + \frac{\sigma\mu^2}{1-\sigma} + \frac{2\tau\sigma\mu}{1-\sigma} \\ &= \frac{2}{n-1} \cdot (\mu + \mu^2). \end{aligned}$$

The last line was obtained by substituting the formulas given earlier for σ and τ and then simplifying. Notice that $\mu + \mu^2 \leq 0$ since $\mu \in [-1, 0]$ as demonstrated above. Thus, $t_2 + t_4 + t_5 \leq 0$.

Summing up, we have

$$\begin{aligned} (\mathbf{x} - \mathbf{c}^{(i+1)}) B_{i+1}^{-1} (\mathbf{x} - \mathbf{c}^{(i+1)}) &= \delta^{-1}(t_1 + t_2 + t_3 + t_4 + t_5 + t_6) \\ &\leq \delta^{-1} \left(1 + \tau^2 + \frac{\sigma\tau^2}{1-\sigma} \right) \\ &\leq 1 \end{aligned}$$

by choice of δ, τ, σ. This concludes the proof that $E_i \cap H \subset E_{i+1}$. ∎

We next turn to the argument that the volumes of the E_i's are shrinking.

LEMMA 6.3. *The following equation is satisfied:*

$$\mathrm{vol}(E_{i+1})/\,\mathrm{vol}(E_i) = \frac{n}{n+1}\left(\frac{n^2}{n^2-1}\right)^{(n-1)/2}$$

PROOF. Let $B_i = LL^T$ be the Cholesky factorization of B_i with positive entries on the diagonal of L. Then $\mathrm{vol}(E_i) = v_n \det(L)$, where v_n is the volume of an n-dimensional sphere and $\det(L)$ is the determinant of L. This follows because E_i is the inverse image of the unit sphere $\{\mathbf{x} : \mathbf{x}^T\mathbf{x} \le 1\}$ under the affine mapping $\mathbf{x} \mapsto L^{-1}(\mathbf{x} - \mathbf{c}^{(i)})$. Thus, we use the standard theorem of calculus that volumes change under affine mappings by the determinant of the affine transformation.

Similarly, $\mathrm{vol}(E_{i+1}) = v_n \det(L')$, where L' is a Cholesky factor of B_{i+1}. We now derive a formula for $\det(L')$:

$$\begin{aligned} L'L'^T &= B_{i+1} \\ &= \delta\left(B_i - \frac{\sigma B_i\mathbf{a}\mathbf{a}^T B_i}{\mathbf{a}^T B_i \mathbf{a}}\right) \\ &= \delta\left(LL^T - \frac{\sigma LL^T\mathbf{a}\mathbf{a}^T LL^T}{\mathbf{a}^T B_i \mathbf{a}}\right) \\ &= \delta L\left(I - \frac{\sigma L^T\mathbf{a}\mathbf{a}^T L}{\mathbf{a}^T LL^T \mathbf{a}}\right)L^T \\ &= \delta L\left(I - \frac{\sigma \mathbf{v}\mathbf{v}^T}{\mathbf{v}^T\mathbf{v}}\right)L^T \end{aligned}$$

where $\mathbf{v} = L^T\mathbf{a}$. Thus,

$$\det(L') = \delta^{n/2}\det(L)\cdot|\det(I - \sigma\mathbf{v}\mathbf{v}^T/(\mathbf{v}^T\mathbf{v}))|^{1/2}.$$

To derive the determinant of $I - \sigma\mathbf{v}\mathbf{v}^T(\mathbf{v}^T\mathbf{v})$, observe that this matrix has one eigenvalue equal to $1-\sigma$ (corresponding to eigenvector $\mathbf{v}$) and $n-1$ eigenvalues equal to 1 (corresponding to eigenvectors forming a basis for the orthogonal complement of $\mathbf{v}$). The determinant is the product of the eigenvalues, i.e., $1-\sigma$. Hence, $\det(L') = \delta^{n/2}(1-\sigma)^{1/2}\det(L)$, so

$$\mathrm{vol}(E_{i+1})/\,\mathrm{vol}(E_i) = \delta^{n/2}(1-\sigma)^{1/2}.$$

Substituting the formulas above for δ and σ gives the result. ∎

To analyze the formula $(n^2/(n^2-1))^{(n-1)/2}n/(n+1)$ occurring in the previous lemma, we use the identity $1 + x \le \exp(x)$ for any real x:

$$\frac{n}{n+1}\left(\frac{n^2}{n^2-1}\right)^{(n-1)/2} = \left(1 - \frac{1}{n+1}\right)\left(1 + \frac{1}{n^2-1}\right)^{(n-1)/2}$$

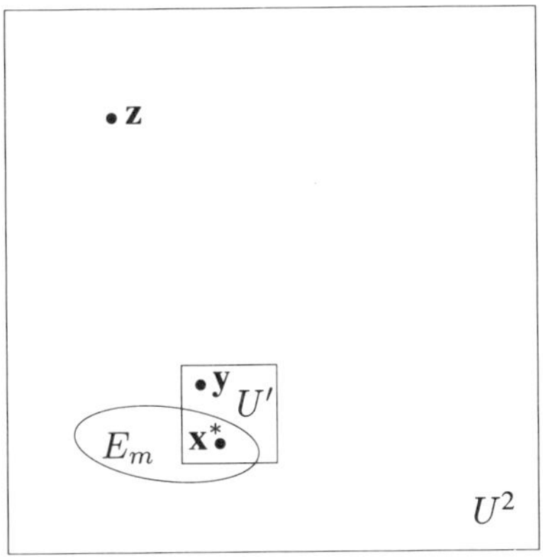

FIG. 6.4. An example of U', $\mathbf{y}$ and $\mathbf{z}$ in the $n = 2$ case.

$$\le \exp\left(-\frac{1}{n+1}\right) \cdot \exp\left(\frac{1}{n^2-1}\right)^{(n-1)/2}$$
$$\le \exp\left(-\frac{1}{n+1} + \left(\frac{1}{n^2-1}\right)\left(\frac{n-1}{2}\right)\right)$$
$$\le \exp\left(-\frac{1}{2(n+1)}\right).$$

We can now complete the description of the ellipsoid algorithm. We continue computing $E_1, E_2, \ldots$ until an ellipsoid of volume less than θ^n is obtained, where θ is defined below. At this point, we take all the values of $f(\mathbf{c}^{(i)})$ stored in list F, and report as minimizer the particular $\mathbf{c}^{(i)}$ encountered during the algorithm with the lowest objective function value. Call this point $\mathbf{c}^*$. Let m be the last value of i in the ellipsoid algorithm.

Let us analyze the difference between $f(\mathbf{c}^*)$ and $f(\mathbf{x}^*)$, where $\mathbf{x}^*$ is the global minimum of f on U^n. This analysis follows Nemirovsky and Yudin. Note that, as observed earlier, $\mathbf{x}^* \in E_i$ for all E_i produced by the algorithm. In particular, $\mathbf{x}^* \in E_m$. Define the affine "compression" map $\phi : \mathbb{R}^n \to \mathbb{R}^n$ as follows:

$$\phi(\mathbf{x}) = (1-\theta)\mathbf{x}^* + \theta\mathbf{x}.$$

This map is bijective, and contracts all points towards $\mathbf{x}^*$. Let $U' = \phi(U^n)$. Observe that U' is a cube of side length θ. Observe also that U' contains $\mathbf{x}^*$ and is contained in U^n. Since $\mathrm{vol}(E_m) < \theta^n$, there is at least one point, say $\mathbf{y}$, in U' but not E_m. Let $\mathbf{z} = \phi^{-1}(\mathbf{y})$; then $\mathbf{z}$ is a point in U^n, and

$$\mathbf{y} = (1-\theta)\mathbf{x}^* + \theta\mathbf{z}. \tag{6.3}$$

See Fig. 6.4 for an illustration of these points and sets in two dimensions.

Now, we observe that since $\mathbf{y} \notin E_m$, there must have been an iteration, say i, such that $\mathbf{y} \in E_i$ but $\mathbf{y} \notin E_{i+1}$. Then consider the halfspace H

constructed by the algorithm such that $E_i \cap H \subset E_{i+1}$. There are two possibilities: Either H was chosen because $\mathbf{c}^{(i)}$ was infeasible, or H was chosen because $\mathbf{c}^{(i)}$ was feasible. The first case is impossible, however, because in this case $U^n \subset H$, and in particular, $\mathbf{y}$ would lie in $E_i \cap H$.

Thus, at step i, $\mathbf{c}^{(i)}$ was feasible. Since $\mathbf{y} \notin E_{i+1}$, we know that $\mathbf{y} \notin H$. Halfspace H, however, contains all points of U^n with objective function better than $\mathbf{c}^{(i)}$, as proved by Theorem 6.3. We conclude that $f(\mathbf{y}) \geq f(\mathbf{c}^{(i)})$. Since $f(\mathbf{c}^*) \leq f(\mathbf{c}^{(i)})$ by choice of $\mathbf{c}^*$, we conclude that $f(\mathbf{y}) \geq f(\mathbf{c}^*)$. Next, by convexity and (6.3),

$$f(\mathbf{y}) \leq (1-\theta) f(\mathbf{x}^*) + \theta f(\mathbf{z}),$$

i.e.,

$$f(\mathbf{y}) - f(\mathbf{x}^*) \leq \theta(f(\mathbf{z}) - f(\mathbf{x}^*)).$$

Now, we claim that $f(\mathbf{z}) - f(\mathbf{x}^*) \leq L$, the Lipschitz bound on f. This follows immediately from the fact that $\|\mathbf{z} - \mathbf{x}^*\|_\infty \leq 1$. Thus,

$$f(\mathbf{y}) - f(\mathbf{x}^*) \leq \theta L.$$

Since $f(\mathbf{c}^*) \leq f(\mathbf{y})$,

$$f(\mathbf{c}^*) - f(\mathbf{x}^*) \leq \theta L.$$

Thus, to have an ϵ-approximate global minimum, we must ensure that $\theta L \leq \epsilon$, i.e., $\theta \leq \epsilon / L$. Recall that θ^n is the volume of E_m, and

$$\operatorname{vol}(E_m) \leq \exp\left(\frac{-m+1}{2(n+1)}\right) \operatorname{vol}(E_1).$$

The formula for E_1 was given; it is an n-dimensional sphere of radius $\sqrt{n/4}$, which has volume $(\pi n/4)^{n/2}/(n/2)!$ if n is even. This is bounded above by $\exp(n \ln n)$. The formula in the case that n is odd is

$$\frac{(n/4)^{n/2} 2^{(n-1)/2} \pi^{(n-3)/2}}{(1 \cdot 3 \cdot 5 \cdots n)},$$

which has the same bound.

Thus, we need to find m so that $\theta \leq \epsilon/L$, i.e., $\operatorname{vol}(E_m)^{1/n} < \epsilon/L$, i.e.,

$$\exp\left(\frac{-m+1}{2n(n+1)} + \ln n\right) < \epsilon/L$$

so

$$\frac{-m+1}{2n(n+1)} + \ln n < \ln(\epsilon/L).$$

This is solved by taking $m \geq 2n(n+1)(\ln(L/\epsilon) + \ln n) + 2$. Recall that m denotes the number of iterations needed by the ellipsoid algorithm. Thus, $O(n^2(\log n + \log(L/\epsilon)))$ iterations are required. Each iteration requires one

Table 6.1. Running times of information-based algorithms.

Type of minimization	**Lipschitz bound**	**Dependence on n**	**Dependence on $\log(L/\epsilon)$**
Nonconvex global	on f	exponential	exponential
Nonconvex local	on ∇f	polynomial	exponential
Convex global	on f	polynomial	polynomial

function evaluation and one gradient evaluation. Unlike the algorithms for global and local minimization presented in the two previous sections, the ellipsoid algorithm requires a significant amount numerical computation in addition to function and gradient evaluation.

This concludes our analysis of the convex optimization case in the information model. The algorithm can be generalized to arbitrary convex constraints, nondifferentiable convex objective functions, and information with some errors. All of these are addressed by Nemirovsky and Yudin. In addition, they discuss worst-case lower bounds and a variety of other algorithms for minimization in the information model.

Hačijan [1979] showed that the ellipsoid algorithm, when specialized to linear programming with data composed of integers, yields a polynomial time algorithm to find the exact optimum. This requires an analysis similar to the various determinant-based bounds of Section 3.2; Hačijan (whose name is often spelled Khachiyan in English) was the first to provide this kind of analysis. As a practical algorithm for linear programming, the ellipsoid method is generally considered inferior to interior point methods. The reason is that the ellipsoid method in general requires a number of steps equal to the worst-case bound, whereas interior point methods usually perform much better than the worst-case bounds.

Table 6.1 summarizes the results of this chapter. In this table, L and ϵ have different meanings for each row. Notice that the convex minimization case achieves the best bounds overall. To achieve this bound requires the sophisticated ellipsoid method.

Exercises

6.1 Suppose $f : U^n \rightarrow \mathbb{R}$ is a *linear* function, that is, a function of the form $f(\mathbf{x}) = \mathbf{c}^T\mathbf{x} + d$. If f is available only as a black box (but is known to be linear), how many function evaluations are required to find a global minimum of f (give lower and upper bounds)?

6.2 Give an example to show that the point found by Algorithm INFO-LOCAL could actually be a local maximum and not a local minimum.

6.3 What is the complexity of finding an ϵ-approximate local minimum of the function $\frac{1}{2}\mathbf{x}^T H\mathbf{x} + \mathbf{c}^T\mathbf{x}$ on U^n using algorithm INFO-LOCAL? The answer should be in terms of n, H, $\mathbf{c}$ and ϵ.

6.4 Let $f : U \to \mathbb{R}$ be a differentiable function of one variable, and suppose f' has a Lipschitz bound of M. Show that an ϵ-approximate local minimum can be found in $O(\log(M/\epsilon))$ function and gradient (derivative) evaluations.

.5 Given $(H, \mathbf{c}, A, \mathbf{b})$ that specify the convex quadratic programming problem of minimizing $\frac{1}{2}\mathbf{x}^T H\mathbf{x} + \mathbf{c}^T\mathbf{x}$ subject to $A\mathbf{x} \geq \mathbf{b}$, show how to find an ellipsoid E_1 guaranteed to contain the minimizer. You should assume that $(H, \mathbf{c}, A, \mathbf{b})$ all have integer entries requiring a total of at most L bits.

.6 Explain how to generalize the ellipsoid algorithm to handle constraints of the form $r_i(\mathbf{x}) \leq 0$ where r_i is a convex differentiable function.

BIBLIOGRAPHY

Alfred V. Aho, John E. Hopcroft and Jeffrey D. Ullman [1974], *The Design and Analysis of Computer Algorithms*, Addison-Wesley, Reading, Mass.

Robert G. Bland, Donald Goldfarb and Michael J. Todd [1981], The ellipsoid method: A survey, *Operations Research* 29:1039–1091.

Lenore Blum, Mike Shub, and Steve Smale [1988], On a theory of computations over the real numbers: NP-completeness, recursive functions and universal machines, *Proc. 29th Symp. Foundations of Computer Science*, 387–397.

Peter Brucker [1984], An $O(n)$ algorithm for quadratic knapsack problems, *Operations Research Letters* 3:163–166.

D. Burton and Ph. L. Toint [1989], On an instance of the inverse shortest paths problem, preprint.

John Canny [1988], Some algebraic and geometric computations in PSPACE, *Proc. 20th Annual ACM Symposium on Theory of Computing,* 460–467.

Bernard Chazelle and Leonidas J. Guibas [1985], Visibility and intersection problems in plane geometry, *ACM Symposium on Computational Geometry*, 135–146.

Alan Cobham [1965], The intrinsic computational difficulty of functions, 24–30 in *Proc. 1964 International Congress for Logic, Methodology and Philosophy of Science*, edited by Yehoshua Bar-Hillel, North-Holland, Amsterdam.

Luis Contesse B. [1980], Une caractérisation complète des minima locaux en programmation quadratique, *Numer. Math.* 34:315–332.

Stephen A. Cook [1971], The complexity of theorem-proving procedures, *Proc. 3rd Annual ACM Symposium on Theory of Computing,* 151–158.

Richard W. Cottle, Steven G. Duvall and Karel Zikan [1986], A Lagrangean relaxation algorithm for the constrained matrix problem, *Naval Research Logistics Quarterly* 33:55–76.

George B. Dantzig [1963], *Linear Programming and Extensions,* Princeton University Press, Princeton, New Jersey.

J. E. Dennis, Jr. and Robert E. Schnabel [1983], *Numerical Methods for Unconstrained Optimization and Nonlinear Equations*, Prentice-Hall, Englewood Cliffs, New Jersey.

Jack Edmonds [1965], Paths, trees, and flowers, *Canad. J. Math* 17:449–467.

Jack Edmonds [1967], Systems of distinct representatives and linear algebra, *J. Res. Nat. Bur. Standards*, 71B:241–245.

R. Fletcher [1987], *Practical Methods of Optimization, Second Edition*, John Wiley and Sons, Chichester.

Michael R. Garey and David S. Johnson [1979], *Computers and Intractability: A Guide to the Theory of NP-Completeness*, W. H. Freeman, San Francisco.

David M. Gay [1979], Computing optimal locally constrained steps, Technical Summary Report 2013, Mathematics Research Center, University of Wisconsin, Madison, Wisconsin.

Philip E. Gill, Walter Murray and Margaret H. Wright [1981], *Practical Optimization*, Academic Press, London.

Gene H. Golub and Charles F. Van Loan [1989], *Matrix Computations, Second Edition*, Johns Hopkins University Press, Baltimore.

C. C. Gonzaga [1989], An algorithm for solving linear programming problems in $O(n^3L)$ operations, in *Progress in Mathematical Programming: Interior-Point and Related Methods*, edited by Nimrod Megiddo, Springer Verlag, New York.

Branko Grünbaum [1967], *Convex Polytopes*, Interscience, London.

L. G. Hačijan [1979], A polynomial algorithm in linear programming, *Dokl. Akad. Nauk SSSR* 244:1093–1096. Translated in *Soviet Math. Dokl.* 20:191–194.

J. Hartmanis and R. E. Stearns [1965], On the computational complexity of algorithms, *Trans. A.M.S.* 117:285–306.

R. Helgason, J. Kennington and H. Lall [1980], A polynomially bounded algorithm for a singly constrained quadratic program, *Math. Progr.* 18:338–343.

Doug Ierardi and Dexter Kozen [1990], Parallel resultant computation, Technical Report 90-1087, Department of Computer Science, Cornell University, Ithaca, New York.

D. B. Iudin and A. S. Nemirovskii [1976], Informational complexity and efficient methods for solving complex extremal problems, *Ekonomika i Matematicheskie Metody*, 12:357–369. Translated in *Matekon: Translations of Russian and East European Math. Economics* 13:25-45, Spring 1977.

Nathan Jacobson [1974], *Basic Algebra I*, W. H. Freeman, San Francisco.

Sanjiv Kapoor and Pravin M. Vaidya [1986], Fast algorithms for convex quadratic programming and multicommodity flows, *Proc. 18th Annual ACM Symposium on Theory of Computing*, 147–159.

N. Karmarkar [1984], A new polynomial-time algorithm for linear programming, *Combinatorica* 4:373–395.

N. Karmarkar [1989], An interior-point approach to NP-complete problems, preprint.

Richard M. Kárp [1972], Reducibility among combinatorial problems, 85–103 in *Complexity of Computer Computations*, edited by Raymond E. Miller and James W. Thatcher, Plenum Press, New York.

V. Klee and G. J. Minty [1972], How good is the simplex algorithm? in *Inequalities III*, edited by O. Shisha. Academic Press, New York.

Donald E. Knuth [1973a], *The Art of Computer Programming, Second Edition, Volume 1, Fundamental Algorithms*, Addison-Wesley, Reading, Mass.

Donald E. Knuth [1973b], *The Art of Computer Programming, Volume 3, Sorting and Searching*, Addison-Wesley, Reading, Mass.

Masakazu Kojima, Shinji Mizuno and Akiko Yoshise [1989], A polynomial-time algorithm for a class of linear complementarity problems, *Math. Progr.* 44:1–26.

M. K. Kozlov, S. P. Tarasov and L. G. Hačijan [1979], Polynomial solvability of convex quadratic programming, *Dokl. Akad. Nauk SSSR* 248:1049–1051. Translated in *Soviet Math. Dokl.* 20:1108–1111.

J. Kuczyński and H. Woźniakowski [1990], Estimating the largest eigenvalue by the power and Lanczos algorithms with a random start, preprint.

Kenneth Levenberg [1944], A method for the solution of certain non-linear problems in least squares, *Quarterly Appl. Math.* 2:164–168.

László Lovász [1986], *An Algorithmic Theory of Numbers, Graphs and Convexity,* Society for Industrial and Applied Mathematics, Philadelphia.

Donald W. Marquardt [1963], An algorithm for least-squares estimation of nonlinear parameters, *J. SIAM* 11:431–441.

Nimrod Megiddo [1984], Linear programming in linear time when the dimension is fixed, *J. ACM* 31:114–127.

M. Mignotte [1982], Some useful bounds, 259–263 in *Computer Algebra: Symbolic and Algebraic Computation* edited by B. Buchberger, G. E. Collins and R. Loos, Springer Verlag, Wien.

S. Mizuno, M. J. Todd, and Y. Ye [1989], Anticipated behavior of path-following algorithms for linear programming, Technical Report 878, School of Operations Research and Industrial Engineering, Cornell University, Ithaca, New York.

Renato D. C. Monteiro and Ilan Adler [1989], Interior path following primal-dual algorithms, Part II: Convex quadratic programming, *Math. Progr.* 44:43–66.

Jorge J. Moré [1978], The Levenberg-Marquardt algorithm: Implementation and theory, 105–116 in *Numerical Analysis: Proceedings of the Biennial Conference, Dundee, 1977*, edited by G. A. Watson, Lecture Notes in Mathematics Vol. 630, Springer Verlag, Berlin.

Jorge J. Moré and Stephen A. Vavasis [1991], On the solution of concave knapsack problems, *Math. Progr.* 49:397–411.

T. S. Motzkin and E. G. Straus [1965], Maxima for graphs and a new proof of a theorem of Turán, *Canad. J. Math.* 17:533–540.

James R. Munkres [1975], *Topology, a First Course,* Prentice-Hall, Englewood Cliffs, New Jersey.

Bruce A. Murtagh and Michael A. Saunders [1987], "MINOS 5.1 user's guide," Technical Report SOL-83-20R, Department of Operations Research, Stanford University, Stanford, CA.

K. G. Murty [1988], *Linear Complementarity, Linear and Nonlinear Programming*, Heldermann Verlag, Berlin.

Katta G. Murty and Santosh N. Kabadi [1987], Some NP-complete problems in quadratic and nonlinear programming, *Math. Progr.* 39:117–129.

A. S. Nemirovsky and D. B. Yudin [1983], *Problem Complexity and Method Efficiency in Optimization*, John Wiley and Sons, Chichester, translated by E. R. Dawson from *Slozhnost' Zadach i Effektivnost' Metodov Optimizatsii*, 1979, Glavnaya redaktsiya fiziko-matematicheskoi literatury, Izdatelstva "Nauka".

Erich Novak [1988], *Deterministic and Stochastic Error Bounds in Numerical Analysis*, Lecture Notes in Mathematics Vol. 1349, Springer Verlag, Berlin.

Christos H. Papadimitriou and Kenneth Steiglitz [1982], *Combinatorial Optimization: Algorithms and Complexity,* Prentice-Hall, New Jersey.

Panos M. Pardalos, Yinyu Ye and Chi-Geun Han [1989], Algorithms for the solution of quadratic knapsack problems, preprint.

P. M. Pardalos and J. B. Rosen [1987], *Constrained Global Optimization: Algorithms and Applications*, Lecture Notes in Computer Science Vol. 268, Springer Verlag, Berlin.

James Renegar [1986], A polynomial-time algorithm, based on Newton's method, for linear programming, Technical Report 07118-86, Mathematical Sciences Research Institute, Berkeley, California.

James Renegar [1989], On the computational complexity and geometry of the first-order theory of the reals, Part I, Technical Report 853, School of Operations Research and Industrial Engineering, Cornell University, Ithaca, New York.

James Renegar and Michael Shub [1988], Simplified complexity analysis for Newton LP methods, Technical Report 807, School of Operations Research and Industrial Engineering, Cornell University, Ithaca, New York.

R. Tyrrell Rockafellar [1970], *Convex Analysis*, Princeton University Press, Princeton, N. J.

Sartaj Sahni [1974], Computationally related problems, *SIAM J. Comp.* 3:262–279.

N. Z. Shor [1970], Convergence rate of the gradient descent method with dilatation of the space, *Kibernetika* 6, no. 2, 80–85. Translated in *Cybernetics* 6, no. 2, 102–108.

K. Sikorski [1984], Optimal solution of nonlinear equations satisfying a Lipschitz condition, *Numer. Math.* 43:225–240.

Steve Smale [1986], Newton's method estimates from data at one point, 185–196 in *The Merging of Disciplines: New Directions in Pure, Applied, and Computational Mathematics* edited by Richard E. Ewing, Kenneth I. Gross, Clyde F. Martin, Springer Verlag, New York.

D. C. Sorenson [1982], Newton's method with a model trust region modification, *SIAM J. Numer. Anal.* 19:409–426.

Éva Tardos [1986], A strongly polynomial algorithm to solve combinatorial linear programs, *Operations Research* 34:250–256.

Aimo Törn and Antanas Žilinskas [1989], *Global Optimization*, Lecture Notes in Computer Science Vol. 350, Springer Verlag, Berlin.

J. F. Traub, G. W. Wasilkowski and H. Woźniakowski [1988], *Information-Based Complexity,* Academic Press, Boston.

J. F. Traub and H. Woźniakowski [1982], Complexity of linear programming, *Oper. Res. Lett.* 1:59-62.

Pravin M. Vaidya [1989], Speeding-up linear programming using fast matrix multiplication (extended abstract), *Proc. 30th Symp. Foundations of Computer Science,* 332–337.

Stephen A. Vavasis [1990a], Quadratic programming is in NP, *Info. Proc. Lett.* 36:73–77.

Stephen A. Vavasis [1990b], Local minima for indefinite quadratic knapsack problems, *Math. Progr.*, to appear.

Stephen A. Vavasis [1990c], *Black-box complexity of local minimization*, Technical Report 90-1132, Department of Computer Science, Cornell University, Ithaca, New York.

Stephen A. Vavasis and Richard Zippel [1990], Proving polynomial-time for sphere-constrained quadratic programming, Technical Report 90-1182, Department of Computer Science, Cornell University, Ithaca, New York.

Yinyu Ye [1988], On the interior algorithms for nonconvex quadratic programming, preprint.

Yinyu Ye [1990], A new complexity result on minimization of a quadratic function over a sphere constraint, preprint.

Yinyu Ye and Edison Tse [1989], An extension of Karmarkar's projective algorithm for convex quadratic programming, *Math. Progr.* 44:157–179.

Richard Zippel [1990], *Algebraic Computation*, manuscript.

INDEX

adaptive, 137, 140
Adler, I., 43
affine transformation, 152
Aho, A., viii, 39, 114

BFGS method, 140
binary search, 39, 116–118, 133
black-box model, 2, 135–156
Bland, R., 148
Blum, L., 34
bounded set, 5
Broyden, C., 140
Brucker, P., 41, 74
Burton, D., 74

Canny, J., 136
Cauchy, A., 96
Chazelle, B., 133
clique problem, 27–29, 81–83
closed set, 4
Cobham, A., 22
combinatorial optimization, 1, 24, 73
compact set, 5, 47
complexity, vii
complexity classes, 18–32
concavity, 108, 109, 118, 119
condition number, 74
Contesse, L., 105
convex programming, 8, 146–155, *see* quadratic programming, convex
convexity, 6–9, 11–15, 121–122, 146–147
Cook, S., vii, 26, 32, 35
Cottle, R., 37

Dantzig, G., vii, 3, 73
decision problem, 18, 35, 41, 77, 100–101
Dennis, J., 2, 135, 140
descent direction, 5, 105–107
Duvall, S., 37

Edmonds, J., 20, 22
ellipsoid method, vii, 43, 57, 146–155

Farkas' Lemma, 9
feasible direction, 5, 105–107, 110, 125
feasible region, 2–3
Fermat's last theorem, 1
first-order condition, 6
Fletcher, R., vii, 1, 10, 122, 140
floating-point numbers, 33

Garey, M., viii, 17, 77
Gay, D., 84
geometric problems, 73
Gill, P., vii
global minimum, 5, 44, 56–62, 136–140
  approximate, 137–138
Goldfarb, D., 140, 148
Golub, G., 4, 19, 33, 34, 87
Gonzaga, C., 42
good approximation, 50
Grünbaum, B., 6
Guibas, L., 133

Hačijan, L., *see* Khachiyan, L.
Han, C., 81
Hartmanis, J., vii, 17
Helgason, R., 37
Hessian, 13, 48

HKL algorithm, 37–42, 74, 120
Hopcroft, J., viii, 39, 114

Ierardi, D., 97
inner product, 4
instance, 17–18
integer programming, 3
interior point method, vii, 42–74, 155
inverse shortest paths problem, 74
Iudin, D., *see* Yudin, D.

Jacobson, N., 95
Johnson, D., viii, 17, 77

Kabadi, S., 103
Kapoor, S., 43
Karmarkar, N., vii, 42, 68, 85
Karp, R., vii, 27
Karush, W., 9
Kennington, J, 37
Khachiyan, L., vii, 43, 57, 155
KKT conditions, 9–12, 15–16, 38, 46, 49, 60, 83, 86, 110, 124
  approximate, 141
Klee, V., 73
Knuth, D., 21, 39, 114
Kojima, M., 43
Kozen, D., 97
Kozlov, M., 43
Kucziński, J., 93
Kuhn, H., 9

Lall, H., 37
least squares, 11, 61
Levenberg, K., 84
line search, 140
linear programming, 3, 11–12, 23–24, 68, 73, 155
Lipschitz condition, 137, 142
local minimum, 5, 103–134, 140–146
  approximate, 140–142
Lovász, L., 2, 6

Marquardt, D., 84
medians, 74
Megiddo, N., 73
Mignotte, M., 96
Minty, G., 73
Mizuno, S., 43, 69
model trust region method, 84, 140
models of computation, 18, 33–34, 135
Monteiro, R., 43
Moré, J., 84, 108
Motzkin, T., 81–82, 103
Munkres, J., 4, 5
Murray, W., vii
Murtagh, B., 3, 103
Murty, K., viii, 1, 77, 103

Nemirovski, A., 136, 140, 146, 153, 155
Newton's method, 49, 52
Novak, E., 140
NP, 22–24, 76, 100
NP-complete, vii, 26–31, 76, 133
NP-hard, 31–32, 35, 101, 103, 141

objective function, 2–3
  polynomial, 136
open set, 4

P, 19, 22, 33, 42, 85, 133
Papadimitriou, C., viii, 9, 17, 34, 100
parameterization, 15, 106–107, 119, 134
Pardalos, P., 81, 100
partitioning algorithms, 100
polynomial number of digits, 20–21, 44–46, 69–72, 77–80, 102
polynomial transformation, 24–25, 27, 35
polynomial-time algorithm, 22, 36–75, 84–85, 107–134, 140
  strongly polynomial, 31, 41, 73
polynomials, 93
  factorization of, 95
  resultant of, 96–98
  rings and, 95
  root separation of, 93, 99
positive definite, 13
positive definiteness problem, 19–21
positive semidefinite, 13

potential function, 47
Powell, M., 140
problem, 17
projected gradient, 143

QKP, *see* quadratic knapsack problem
QP, *see* quadratic programming
quadratic constraint, 35, 100
quadratic function, 2, 14
quadratic knapsack problem (QKP)
  convex, 37–42, 74
  local minima, 107–134
  nonconvex, 81
quadratic programming (QP), 3, 31–32
  box constraints, 80–81, 101
  convex, 36–75
  enumerative algorithms, 99–100
  knapsack, *see* quadratic knapsack problem
  nonconvex, 44–46, 76–104
  simplex constraints, 81–83, 134
  sphere constraint, 84–95, 101–102

randomized algorithms, 140
real numbers, 19, 33–34
real-number model, 33–34, 73, 135–156
Renegar, J., 42, 51, 68, 136
Renegar-Shub method, *see* interior point method
Rockafellar, R., 7
Rosen, J., 100

Sahni, S., 32, 76
satisfiability problem (SAT), 25–26, 31
Saunders, M., 3, 103
Schnabel, R., 2, 135, 140
second-order conditions, 6, 86–87, 104–107
Shanno, D., 140
Shor, N., 146
Shub, M., 34, 42, 51, 68
Sikorski, K., 138
simplex method, vii, 3, 73
singular value decomposition, 53
Smale, S., 34, 51
Sorenson, D., 84
sorting, 39, 114
standard form, 74
Stearns, R., vii, 17
Steiglitz, K., viii, 9, 17, 34, 100
Straus, E., 81, 82, 103
subdifferential, 122, 146
subset-sum problem, 29–31, 81
Sylvester matrix, 97

Tarasov, S., 43
Tardos, É., 73
Todd, M., viii, 69, 148
Toint, Ph., 74
topology, 4–5
Törn, A., 135
Traub, J., 73, 135, 140
Tse, E., 43
Tucker, A., 9
Turing machine, vii, 18–34

Ullman, J., viii, 39, 114
unbounded optimization problem, 5, 17, 75, 77–80

Vaidya, P., 43, 68
Van Loan, C., 4, 19, 33, 34, 87
Vavasis, S., 76, 85, 101, 108, 119, 141, 145

Wasilkowski, G., 135, 140
Woźniakowski, H., 73, 93, 135, 140
Wright, M., vii

Ye, Y., 43, 69, 81, 85
Yoshise, A., 43
Yudin, D., 136, 140, 146, 153, 155

Zikan, K., 37
Žilinskas, 135
Zippel, R., 85, 95, 101